CORPORATE SOCIAL RESPONSIBILITY AND GLOBALISATION

AN ACTION PLAN FOR BUSINESS

JACQUELINE CRAMER

Greenleaf
PUBLISHING
2006

© 2006 Greenleaf Publishing Ltd

Published by Greenleaf Publishing Limited
Aizlewood's Mill
Nursery Street
Sheffield S3 8GG
UK
www.greenleaf-publishing.com

Printed in Great Britain on acid-free paper by Antony Rowe Ltd, Chippenham, Wiltshire.
Cover by LaliAbril.com.
Cartoons: Jan Dirk Barreveld / Comic House NL

British Library Cataloguing in Publication Data:
A catalogue record for this book is available from the British Library.

ISBN-10: 1-874719-31-4
ISBN-13: 978-1-874719-31-1

Contents

List of boxes, tables and figures

Preface

This book was written in close co-operation with 20 companies that took part in the 'Corporate Social Responsibility in an International Context' programme. This programme was run by MVO Nederland (CSR Netherlands) (www.mvonederland.nl) (previously by the National Initiative for Sustainable Development). It ran from the beginning of 2003 until the end of 2005 and was financed by the Dutch Ministry of Housing and the Environment.

The following business representatives contributed to this book: Onno Mulder, Dennis de Jong and Mark Hanhart, ABN AMRO; Werner Buck, Friesland Foods; Maarten Smits and Peter Burger, Fugro; René Zijlstra, Royal Haskoning; Mark van Rijn, Heineken; Pety Versendaal, Josbert Kester and Marcel Ririassa, Pentascope; Bert Fokkema, Monique de Wit and Tim van Kooten, Shell; Rob de Ruiter and Dirk den Ottelander, Thermphos; Marlotte Herweijer and Timo de Grefte, Koninklijke Wessanen; Mark Diepstraten, Koninklijke Houthandel Wijma; Dave Boselie, AgroFair; Caspar Vedral, AXA Stenman; Heleen Hooij and Kees Lansbergen, Bloemen-veiling Aalsmeer; Theo de Groot, MPS; Pieter Huygens and Edo Beukema, De Bijenkorf; Sabine Hulsman and Karen Liedenbaum, The Cookie Company; Viviënne van Eijkelenborg, Difrax; George

Tiemstra, Koninklijke KPN; Monique Berings, Merison; Jan Neeleman and Albert Getkate, Rijnvallei; and Paula Koelemij, Simon Lévelt.

The following students were also involved: Moniek Dotinga, Floor Klein, Floor Overbeeke, Rosalie Porte, Tessa Roorda and Kim Teeselink and the postgraduate students Eddy van Hemelrijck and Esther Schouten.

An editing committee, consisting of Bernedine Bos, Mark Diepstraten, Bert Fokkema, Paula Koelemij, Onno Mulder and Johan Wempe, took a critical look at the manuscript.

Two colleagues from MVO Nederland were actively engaged during the course of the programme: Bernedine Bos and Annelies van der Veen. They formed the supporting programme team together with the author of the book.

Without the help of all the people mentioned above, this book could not have been written. Therefore the author wishes to thank all those involved in the process of writing this book.

1

Corporate social responsibility: a global challenge for business

Companies that operate globally are being increasingly asked to act in a socially responsible manner. In addition to achieving financial returns, they are also expected to care for the environment, their employees and the local community. In order to make sure companies observe certain rules of conduct, various international institutions have drawn up guidelines and standards during the last few decades. Observance of these rules of conduct is gaining an ever more important place on the company agenda. A company cannot permit itself to be publicly criticised because of poor working conditions, environmental scandals or a violation of human rights. Such criticism damages a company's reputation, which may result in a fall in sales figures and employees becoming demotivated. Some companies have turned this threat into an opportunity and now present themselves as socially responsible companies. By being a socially responsible enterprise, they try to increase their market share, innovative power and staff motivation to work for the company. Moreover, they try to achieve cost advantages, while simultaneously shaping their own moral responsibility. This ambition is called corporate social responsibility.

Companies that embrace the concept of corporate social responsibility do not wait until the government imposes particular rules or laws. They look ahead and determine for themselves which environmental and social measures they are able or willing to take. They choose those measures that fit in with their own vision and business strategy. But they also take account of what the outside world asks of them. They have developed an identity that is based on finding a responsible balance between **people** ('social well-being'), **planet** ('ecological quality') and **profit** ('economic prosperity') (see Box 1.1). Their attention shifts from gaining purely financial profit to sustainable profit. And finally they want to communicate openly about these things with their employees and with the diverse societal groups that directly or indirectly have a stake in their company (the 'stakeholders'). That, in a nutshell, is what corporate social responsibility means (Cramer 2003).

Thus, the choices a company makes concerning people, planet and profit depend on the company's vision and strategy, but it is also important to take what the outside world expects from the company into consideration. This can create tension, particularly in an international context, because not everybody has the same attitude towards problems such as human rights, integrity and care for the environment. Nevertheless, every self-respecting international company has to know how to deal with such dilemmas.

Corporate social responsibility in an international context is not only high on the agenda of large, multinational companies. As a result of economic globalisation, smaller companies are also becoming more involved in a network of international suppliers and customers. Within this network, they are increasingly held accountable for certain activities in their product chains.

However, companies wishing to run an international, sustainable business encounter all sorts of problem. For example:

People relates to a range of subjects, including both internal and external social policies. This concerns not only the company's own part in the chain, but also the rest of the chain for which the company can take responsibility.

Internal social policy includes the nature of employment, labour/management relationships, health and safety, training and education, as well as diversity and opportunities.

External social policy encompasses three main categories, each with several sub-categories, namely:

1. Human rights issues, including strategy and management, non-discrimination, freedom of association and collective bargaining, child labour, forced and compulsory labour, disciplinary practices, security practices and indigenous rights

2. Society, including community activities, bribery and corruption, financial contributions to political parties, competition and pricing

3. Product responsibility, including consumer health and safety, products and services, advertising and respect for personal privacy

Planet relates to the environmental impact of the company's production activities: the use of scarce goods (such as energy, water and other raw materials) and the environmental impact in the product chain.

Profit stands for the company's contribution to economic prosperity in the broadest sense. Here, a distinction is made between direct and indirect economic impact. Direct impact involves the monetary flows between the organisation and its key stakeholders and the impact which the organisation has on the economic circumstances of those stakeholders.

The indirect impact is related to the spin-off from company activities in terms of innovation, the contribution of the sector to gross domestic product or national competitiveness and the local community's dependence on the company's activities.

Box 1.1 The three pillars of corporate social responsibility are people, planet and profit

- How can companies find their way through the maze of guidelines and standards that international organisations have drawn up with regard to corporate social responsibility?

- How can they deal with the tension between observing the international code of conduct with regard to human rights, integrity and environmental policy on the one hand and local attitudes towards these themes on the other? Which rules and procedures can they use?

- To what extent does the interpretation of corporate social responsibility depend on the political culture of a certain country?

- How can they organise chain responsibility in the international product chain(s) in which they operate and how far-reaching is this responsibility?

- What contribution is expected from companies that operate internationally concerning the fight against poverty and strengthening the local economy of developing countries?

- What is the future for corporate social responsibility in an international context?

This book helps companies to find answers to these questions. It is based on the experiences of 20 large and smaller companies, which operate globally. Almost all of these companies have their head office in the Netherlands (see Appendix 1). The book is mainly aimed at companies wishing to be socially responsible with respect to their international activities, but which are not sure how to go about this. It is of particular interest to two groups of companies.

First, the book has been written for companies with branches in foreign countries and which are confronted with differences in political culture, morals and legislation concerning social and environmental policies. They must steer between internationally accepted guidelines and standards and what is accepted locally. Traditionally, it has mainly been international companies that have had to deal with these problems. However, as an increasing number of small and medium-sized companies now also cross international borders, it is also becoming a problem for this group of companies.

Second, the book is intended for companies wishing to be socially responsible in the international product chain(s) in which they operate. They are faced with the question of how to organise this chain responsibility and whom they must involve in it. As a result of the continuing internationalisation of product chains, these problems play a role in both large and smaller companies.

The book offers concrete guidelines, step-by-step plans and practical examples of company experiences and provides an insight into the importance of the **corporate social responsibility in an international context** subject. Using this book, companies can practise corporate social responsibility in an international context. Corporate social responsibility in a global world is therefore not a threat, but rather a challenge for business.

2
Observing international rules of conduct

2.1 A maze of international rules of conduct

The number of international rules of conduct with regard to corporate social responsibility has grown steadily, particularly since the end of the Second World War. The creation of these agreements has been linked to the globalisation process, which has become increasingly evident.

The first generation of international agreements were mostly decided on during the 1970s (Jenkins *et al.* 2002). These agreements were mainly intended to support governments in developing countries to control multinational companies that operated on their soil. Since the local legislation of these countries offered little footing, an international code of conduct was very helpful in those days.

The second generation of international agreements stems from the 1990s. Privatisation, free-trade agreements and continuing economic

integration led to a change in attitude on the part of the governments of developing countries. Instead of controlling multinational companies, there was fierce competition between countries to attract such companies. At the same time, globalisation led to increased attention to corporate social responsibility. This was linked to a shift in power between national states, businesses and citizens. The market's influence was increased as a result of globalisation of economic activities. More power meant greater responsibility. Citizens now also had more opportunities to exert their influence, partly as a result of the emergence of digital technology, which made it easier for them to communicate with each other worldwide and to expose scandals via the media. As a result, companies were forced to take the wishes and demands of the different stakeholders into consideration more often. Openness and transparency became keywords.

Internationally operating companies could not permit themselves to be criticised by the public with regard to child labour, terrible working conditions or environmental scandals. This was harmful to their reputation and gave no proof of corporate social responsibility. In order to ensure that companies observed certain rules, various international institutions, trade organisations, multi-stakeholder organisations and even individual companies have drawn up guidelines and standards concerning corporate social responsibility. A guideline contains guiding principles, while a standard states the output which is expected from companies.

All the guidelines and standards that have been drawn up are intended to offer guidance when giving shape to corporate social responsibility. Most of these rules of conduct are not legally binding, but they contain a moral obligation to act accordingly and can be disciplinarily enforced within a company. The intention and contents of these agreements do not differ greatly but, in practice, this maze of guidelines and standards often appears to be counterproductive, as people are no longer able to see the wood for the trees.

To create some order, a diagrammatical overview of the most important guidelines and standards has been made (see page 26). Based on this, a plan of action has been drawn up. Next, company examples will be used to illustrate that the further elaboration of this plan of action should be attuned to the specific context of the company at stake.

2.2 Plan of action

The plan of action consists of two actions, which are explained below.

To start with, companies are recommended to obtain an initial impression of their current situation regarding corporate social responsibility in an international context within their own company. This zero-assessment does not need to be too detailed. A rough assessment will suffice. After reviewing the available guidelines and standards, the OECD *Guidelines for Multinational Enterprises* (2000) turned out to be the best guidelines to start with. The UN Global Compact is a useful addition to this (www.unglobalcompact.org). Both are all-encompassing guidelines, which show many similarities. Action 1, therefore, is as follows:

> **Action 1**: Assess the business's current state of affairs regarding corporate social responsibility in an international context based on the OECD guidelines and determine priorities and a code of conduct based on this. Communicate these priorities and the code of conduct to the stakeholders.

The OECD (Organisation for Economic Co-operation and Development) guidelines are subscribed to by governments of the OECD member states and a growing number of other countries were added as a result of co-operation between the business world, the trade union movement and other social organisations. Governments subscribing to the OECD guidelines are expected to encourage companies to give shape to their activities both at home and abroad in a socially responsible way. The most important recommendations for companies, stated in the OECD guidelines, are summarised in Box 2.1.

The OECD guidelines can easily be transformed into a questionnaire. The objective of doing so is to obtain the opinions of relevant people within your organisation concerning the current situation with regard to corporate social responsibility. Their answers will give you an overview of the areas where your company already performs well, crucial areas where it performs reasonably well (white spots) and areas where it performs reasonably well, but which are not or only slightly risky. Based on this, the company's management can determine the policy priorities with regard to corporate social responsibility in an international context and draw up a code of conduct. It is a good idea to present these policy priorities and the code of conduct to the stakeholders who are important to the company at an early stage. The stakeholder expectations and demands can then be included when developing the policy for corporate social responsibility.

In order to elaborate this policy Action 2 will be helpful:

> **Action 2**: Develop the policy priorities using theme-specific, international guidelines and standards concerning corporate social responsibility.

GENERAL POLICIES

Enterprises should contribute to economic, social and environmental progress with a view to achieving sustainable development; respect the human rights of those affected by their activities; encourage the creation of local capacity through close co-operation with the local community; encourage, where possible, business partners, including suppliers and subcontractors, to apply principles of corporate conduct compatible with the guidelines; abstain from any improper involvement in local political activities.

DISCLOSURE

Enterprises should disclose timely, regular, reliable and relevant information regarding their activities.

EMPLOYMENT AND INDUSTRIAL RELATIONS

Enterprises should respect the right of their employees to be represented by trade unions and other bona fide employee representatives. Moreover, they should not discriminate against their employees and they should contribute to the effective abolition of child labour and all forms of forced or compulsory labour.

ENVIRONMENT

Enterprises should take due account of the need to protect the environment, public health and safety; establish and maintain a system of environmental management appropriate to the enterprise and provide adequate education and training to employees in environmental, health and safety matters.

COMBATING BRIBERY

Enterprises should not, directly or indirectly, offer, promise, give or demand a bribe or other undue advantage to obtain or retain business or other improper advantage. Nor should enterprises be solicited or expected to render a bribe or other undue advantage.

Box 2.1 *OECD Guidelines for Multinational Enterprises*: the main recommendations (continued opposite)

CONSUMER INTERESTS

When dealing with consumers, enterprises should act in accordance with the business, marketing and advertising practices and should take all reasonable steps to ensure the safety and quality of the goods and services they provide.

SCIENCE AND TECHNOLOGY

Enterprises should adopt, where possible in the course of their business activities, practices that permit the transfer and rapid diffusion of technologies and know-how, with due regard for the protection of intellectual property rights.

COMPETITION

Enterprises should refrain from entering into or carrying out anti-competitive agreements among competitors.

TAXATION

Enterprises should contribute to the public finances of host countries by making timely tax liability payments.

Box 2.1 (from previous page)

Various theme-specific, international guidelines and standards are useful when developing the policy priorities selected in Action 1, such as guidelines and standards with regard to human rights, labour rights, the environment and corruption. They can serve as a guide for improving and safeguarding company performance regarding specific themes. These guidelines and standards can vary greatly in terms of content, scope and stakeholder involvement. The guidelines and standards have been classified into different categories in Box 2.2.

CONTENT

- Performance-oriented standards and guidelines: minimum standards and guidelines regarding the performance a company should achieve in an economic, ecological and social sense
- Process-oriented standards and guidelines: procedures that a company should follow in shaping corporate social responsibility

Together these procedures form the management system.

STAKEHOLDER INVOLVEMENT

- Unilateral: developed by the company itself
- Bilateral: developed by two parties (e.g. a company and a trade union)
- Multilateral or stakeholder-oriented: developed by a network of organisations based on extensive negotiations

STANDARD OR GUIDELINE RANGE

- Generic standards or guidelines: these cover all themes relevant to corporate social responsibility
- Theme-specific standards or guidelines: these focus on specific themes, such as labour, environment, corruption and human rights
- Standards and guidelines that focus on specific stakeholders/ target groups, on specific sectors and/or on specific regions

Box 2.2 Standards and guidelines: classification in categories

Source: Leipziger 2003

The most important theme-specific guidelines and standards are listed in Box 2.3.[1]

The current situation is as follows:

- For internal social policies (particularly labour norms), generally accepted performance- and process-oriented standards (including a framework for certification) are available

- For external social policies regarding human rights and corruption, performance-oriented standards and guidelines are generally accepted, while process-oriented guidelines also exist for anti-corruption

- For external social policies focused on ethical behaviour, a process-oriented, but no generally accepted performance-oriented standard, is available

- For the remaining issues concerning external social policies (particularly local community involvement), no performance-oriented or process-oriented standards or guidelines exist

- For environmental policies, generally accepted performance-oriented guidelines and process-oriented standards exist. The performance-oriented standards already available are sector- and region-specific or based on multi-stakeholder agreements

- For economic policies (the company's contribution to economic prosperity in the broadest sense), standards are (still) lacking.

As a company makes progress in implementing corporate social responsibility, the package of theme-specific guidelines and standards to be applied will gradually increase and become an integrated entity.

---

1 Please refer to Appendix 2 and to Leipziger 2003 for a more detailed explanation.

HUMAN RIGHTS

- The Universal Declaration of Human Rights (UN framework for human rights); performance-oriented standard

LABOUR RIGHTS

- The International Labour Organisation: Tripartite Declaration of Principles concerning Multinational Enterprises and Social Policy (integration of the main ILO conventions and recommendations); performance-oriented standard

- Social Accountability 8000 (worldwide, commonly used standard in the area of labour norms); performance- and process-oriented standard (including certification options and certifiable standard)

- OHSAS 18001 (standard for health and safety policy); process-oriented standard

- AccountAbility 1000 Framework (standard for social and ethical accounting); process-oriented standard

ENVIRONMENT

- The Rio Declaration on Environment and Development (UN starting points regarding environment and development); performance-oriented guideline

- The CERES (Coalition for Environmentally Responsible Economies) principles (10 principles that cover the main environmental issues and also form the basis for the GRI guidelines [see below]); performance-oriented and to a lesser extent process-oriented principles. A comparable set of principles is The Natural Step.

- ISO 14001; process-oriented standard for an environmental management system

CORRUPTION

- The OECD Convention on Combating Bribery of Foreign Public Officials in International Business Transactions (legal framework for coping with corruption practices); legally binding performance standard

Box 2.3 The main theme-specific guidelines and standards
(continued opposite)

- The Business Principles for Countering Bribery (multi-stakeholder framework concerning corruption); performance- and process-oriented guidelines

ECONOMY

Standards related to the company's contribution to economic welfare in the broadest sense are lacking.

GENERIC

In addition to the OECD guidelines for multinational enterprises and the UN Global Compact, as mentioned previously, the following generic guidelines may also be helpful:

- IFC guidelines of the World Bank (criteria for financing social and environmental policy projects); performance-oriented standard
- The 'Sustainability: Integrated Guidelines for Management' (SIGMA) project (attempt to integrate all aspects of corporate social responsibility in one framework); process-oriented standard (not yet generally accepted; still under preparation)
- Global Reporting Initiative (guidelines for the standardisation of reporting methods for corporate social responsibility; based on the main guidelines and standards; increasingly accepted as the basis for reporting); process-oriented guidelines

Box 2.3 (from previous page)

However, the exact interpretation of corporate social responsibility and, therefore, the use of guidelines and standards will depend on the specific local situation of each country: the culture, the social and political context, the government's attitude and the most urgent social problems (see Chapters 3 and 4).

2.3 The use of international rules of conduct at Fugro, Friesland Foods, Royal Haskoning and Koninklijke Wessanen

The elaboration of the plan of action described above can differ between companies. This is illustrated below based on company examples.

2.3.1 Example: Fugro

Fugro is an international company that collects, processes and interprets data about the Earth's surface and the seabed. The company has a highly decentralised organisation, with branches in more than 50 countries and more than 8,000 employees. Fugro carries out activities that have very few risks with regard to corporate social responsibility. Nevertheless, the company wanted to be sure it was following the OECD guidelines. Klaas Wester, CEO of Fugro, explained this as follows:

> Observing the OECD guidelines is not an objective in itself for Fugro. It is the logical result of the management's general ambition to run a financially successful company without harming the environment and, at the same time, guaranteeing a safe and pleasant working environment.

At the beginning of 2004, Fugro started to test the company's policy within the worldwide organisation against the OECD guidelines. Peter Burger, the co-ordinator for corporate social responsibility at Fugro at that time, produced a questionnaire based on all the points in the OECD guidelines. An empty box was placed after each point and the local manager had to indicate whether the local policy was fully in line with the OECD guideline concerned, to some extent in line the OECD guideline concerned or not in line with the OECD guideline concerned. There was also the possibility of choosing 'not familiar with' or 'not applicable'. Since the local managers were often unfamiliar with the OECD guidelines, the questionnaire was made as simple and as straightforward as possible.

In order to ensure that the objective of the questionnaire was clear in advance, Burger explained it during an international gathering of 80 managers from different parts of the world. Thirty-five subsidiary companies were then asked to complete the questionnaire. The reactions showed that Fugro was already advanced in some areas (such as corruption and the environment, as well as health and safety), while other aspects of corporate social responsibility had been given less attention (for example, supplier performance regarding corporate social responsibility and the selection of local agents). In general, though, the results were not bad.

Burger was confronted with cultural differences. Some people were apprehensive about being open, as was the case in, for example, the United States, where only legal activities are reported on. There is a fear of making matters public which might create legal liabilities or which might affect the stock price. The differences in dealing with trade unions also played a role.

Based on the results of the questionnaire, Fugro determined priority areas where it wanted to improve policy. For example, a code of conduct, derived from the OECD guidelines, was being drawn up. Observance of this code of conduct will be guaranteed by a monitoring system. Reporting and embedding the code of conduct will be included in the subsidiaries' management systems. More attention will also be given to communicating Fugro's policy with regard to corporate social responsibility to the stakeholders. 'The intention here will be to not pretend that Fugro does more than it does or can live up to. As a service company, Fugro's contribution to corporate social responsibility is less obvious than that of a production company,' Maarten Smits, Peter Burger's successor, states.

2.3.2 Example: Friesland Foods

Friesland Foods is a multinational company concerned with the development, production and sale of a wide range of dairy products and fruit drinks. Friesland Foods' brands are strongly represented in the dairy markets, particularly in Western and Central Europe, the Middle East, West Africa and South-East Asia. It has 90 branches and 17,500 employees, 12,000 of whom work outside the Netherlands.

Friesland Foods has also used the OECD guidelines to map its existing activities related to corporate social responsibility. The inventory, which was completed in 2001, showed the following:

Friesland Foods meets the OECD guidelines. This is certainly accurate for the company's activities in the Netherlands, but also for those activities carried out by branches in other countries. However, the conditions for combating corruption require a clearer frame of reference for Friesland Foods. Friesland Foods has also changed the right of trade unions to exist to officially recognised trade unions.

With regard to its activities in the Netherlands, the company does more than the OECD guidelines require in the following areas:

- Employment/industrial relations

- The environment

- Consumer/customer interests

For the various branches, this is particularly the case for the following area:

- Employment/industrial relations

The environment certainly receives attention in the foreign branches, but is not completely in accordance with the OECD guidelines.

Friesland Foods' behaviour with regard to corporate social responsibility is insufficiently recorded within the organisation. Increased openness and transparency to the outside world is also desirable. It is noticeable that the branches outside Western Europe are more active in communicating their activities regarding corporate social responsibility than the branches of Friesland Foods in Western Europe. In Western Europe, many of the corporate citizenship matters are considered normal, while this is certainly not the case elsewhere.

Based on the results of the aforementioned inventory, it was decided to draw up a code of conduct for the company, to serve as a guideline for the actions of the management and employees. It contains the values, rules of behaviour and guidelines which Friesland Foods takes into consideration in its relationships with consumers, customers, shareholders, employees, business partners and the communities in which Friesland Foods is active. The draft code of conduct was introduced to the company's 'top 100 management' in the middle of 2002. It was agreed to gain experience with the draft code for a period of nine months, to look at any practical experiences and dilemmas and to then decide whether the draft code had be altered or not. After a few slight alterations, the code of conduct was formally introduced in the summer of 2003.

In the middle of 2004, it was agreed that an annual compliance statement would be signed by the top 100 management. By signing this compliance statement, the managers declare to have acted in accordance with the guidelines stated in the code of conduct. The compliance statement also allows space to indicate the areas where compliance was not possible. The manager concerned must then state in which way and within which time-frame he or she will be able to comply.

In the middle of 2005, a whistle-blower policy was introduced to the top 100 management. The procedural implementation took place in the autumn of 2005, after which the policy will be tested and evaluated at the end of 2006.

In addition to drawing up a code of conduct, Friesland Foods distributed a positioning document, based on the inventory of existing activities, within the company and also made it available via the Friesland Foods website. Friesland Foods uses this document to explain its responsibilities and basic principles concerning corporate social responsibility. The points of attention that are explained in this positioning document under the heading 'Corporate Citizenship' are the personnel policy, the environment, corporate social responsibility, farm milk and drinking water programmes, food safety, infant formula, food and health, modern biotechnology, genetically modified organisms and the code of conduct. A level of ambition and related actions are given for each area of attention. During the annual top 100 management meeting in 2005, colleagues from various countries presented concrete initiatives for these areas of attention, which was very stimulating.

Looking back at Friesland Foods' experiences with applying the OECD guidelines, Werner Buck, the co-ordinator for corporate social responsibility within the company, concludes the following:

- The OECD guidelines are useful for assessing where an international company stands with regard to corporate social responsibility in an international context

- A top-down approach is often inevitable, but requires a support base within the organisation. This can be created by first testing the draft policy and procedures and then including the experiences of the managers and employees in the evaluation and definitive version, as well as for the introduction of the policy and procedures

- A step-by-step approach is very important in order to fulfil the requirements for corporate social responsibility. Commitment from the group's management and a number of key figures is crucial in this regard

2.3.3 Example: Royal Haskoning

Royal Haskoning is the oldest Dutch engineering consultancy, founded in 1881. In addition to employing engineers, the company now also employs consultants and architects. It employs 3,000 people, 1,600 of whom are located in the Netherlands. It has 53 branches, of which 39 are in various countries outside the Netherlands.

Royal Haskoning also started the process of corporate social responsibility using the OECD guidelines. The co-ordinator for the process, René Zijlstra, however, decided not to make a questionnaire out of the OECD guidelines that could be completed by the people responsible. Zijlstra explains why:

> Royal Haskoning is reasonably centrally controlled and has a dominant cross-border division structure. In view of the emphasis on co-operation and harmonisation, policies are often decided on in meetings of the Divisional Directors and the Board of Directors, known as the Management Council. As a result of this method of management and this way of deciding policies, it was decided that the Management Council would discuss the issue instead of sending round a questionnaire.

During the discussion, all the areas of attention from the OECD guidelines were checked to see what their relevance was for Royal Haskoning. Zijlstra also conducted a provisional assessment of the degree to which the organisation operated in accordance with the guidelines. That made the work expected from the Management Council tangible and showed the points where improvement was possible.

A team of people within the organisation then grouped the activities of Royal Haskoning with regard to corporate social responsibility around four pillars: (1) core values and policy; (2) products and services; (3) operational management; and (4) social involvement. This division was more familiar for the company than the usual division of the three Ps (people, planet and profit). This division was combined with the areas of attention from the OECD guidelines to form a matrix, where the four pillars were given along the horizontal axis and the

OECD areas of attention were given along the vertical axis. This matrix formed the starting point for a brainstorming session between 20 people from the company with an affinity for the corporate social responsibility theme. Together, they selected the ten priorities that required most attention. The selection criteria were: possible business risks, a support base within the organisation, the division across the four pillars and the possibility of developing the priorities step by step.

Some examples of the selected priorities are: implementing ISO 14001, formulating a sponsoring policy, including a social report in the general annual report and setting up a business integrity management system in accordance with the standard of the international trade organisation FIDIC to which Haskoning is affiliated. The Management Council approved the selected priorities and their further development.

René Zijlstra is positive about the process that has taken place within Royal Haskoning based on the OECD guidelines. In his words:

> It formed an excellent starting document for unravelling the concept of corporate social responsibility and to translate it into company-specific circumstances. The guidelines can also be used when performing a type of zero-assessment of the state of affairs within the company, but such a zero-assessment must be in keeping with the company-specific situation. We have, for example, not chosen to send out a questionnaire, but to discuss matters in the Management Council. In order to gain a support base within the organisation, it is important, when giving shape to corporate social responsibility, to remain as close as possible to the company's daily practices.

2.3.4 Example: Koninklijke Wessanen

Koninklijke Wessanen (Royal Wessanen) is a Dutch multinational food business group with 9,000 employees. The company is active in the geographical regions of Europe and North America. The focus is placed on health (healthy and natural food products, such as Zonnatura) and premium-taste food products; quality and authenticity are of particular importance.

Disappointing economic results prompted the company to make radical organisational changes and cost savings in 2002–2003. In 2003, Ad Veenhof became the new CEO. Despite the financially difficult situation, Veenhof wanted to continue to give attention to the corporate social responsibility theme. From his (Roman Catholic) background, it is natural for him to think and talk about themes such as corporate social responsibility. He explains this as follows:

> For me, corporate social responsibility involves tackling three important issues. The first is that our planet is changing at an alarmingly fast pace. The second is the lack of transparency in, for example, the food chain. The third is that we are becoming increasingly more individualistic with our freedom, while, at the same time, certain responsibilities are more and more often passed on to the government. In other words, we, as a society, do not look at ourselves in the mirror often enough.

According to Veenhof, a company can most certainly influence these issues, partly via its products, but also by striving for more transparency within the chain. 'In the past, initiatives with regard to corporate social responsibility were taken within Wessanen, but the attention for it waned during a period of reorganisation. Therefore, our first initial concern is to produce solid foundations,' said Veenhof.

The process was started in February 2004 with a meeting of a small project group consisting of the corporate communications staff director, the corporate quality staff director and an employee from the corporate quality department, who was also responsible for co-ordinating the project. It was agreed to take the OECD guidelines as a starting point for Wessanen's policy on corporate social responsibility. In order to gain insight into the state of affairs concerning corporate social responsibility, telephone interviews were held with the managing directors of the Wessanen companies. The interview questions were based on the OECD guidelines, but adapted to the situation within Wessanen and this led to a shortlist of 15 questions. These questions were sent to the managing directors before the interview, together with a summary of the aspects of corporate social responsibility that could play a role in operational management. This summary was based on the themes from the Global Reporting Initiative (GRI) guidelines.

The companies interviewed are a varied selection of European and North American subsidiaries, production companies, distribution companies and marketing companies, health food products and premium-taste food products. In this way, corporate social responsibility could be approached from the different company interests. The investigation quickly showed that the different Wessanen companies dealt with corporate social responsibility very differently.

In theory, Wessanen is a front-runner with regard to corporate social responsibility where its product portfolio is concerned, but the theme was not well known by local management and the subjects considered important varied greatly by location. Formulating, implementing and reporting a policy of corporate social responsibility and sustainable activities was fragmentary and there were large differences in how this was done at each location. The influence that the individual companies could exert on the chain also varied greatly.

The above experiences underlined the need to develop a central vision on corporate social responsibility and to then have this implemented by the bodies that had to work with it. The process was started and the first step was to formulate a code of conduct and policy priorities with regard to corporate social responsibility.

The code of conduct contains the mission statement, the core values, the company's basic principles and the rules for certain matters (such as fraud and promotional gifts). This code has been drawn up and implemented from the top down.

Next, the priorities for the policy on corporate social responsibility were determined by the members of the project group. The results of the telephone interviews were used as input and the priorities were presented to the group's management team afterwards.

This resulted in the selection of seven themes which must, in the next few years, give direction to the policy. Examples of the way in which these themes are given shape within Wessanen are provided in Table 2.1.

The responsibility for corporate social responsibility lies with the corporate quality department. The objectives regarding corporate social responsibility will, from now on, be included in the 'Balanced Scorecard' of this department and will, therefore, become part of strategic planning. In doing so, corporate social responsibility will have a strong position at central level.

Two actions were defined for 2005:

1. Setting up a corporate social responsibility report (including defining indicators and setting up a report structure)

2. Implementing ISO 9001 and ISO 14001 for better control and management of the process

These actions must make it possible for the initiative to be placed with the companies in the future.

Wessanen's experience with using the OECD guidelines as a starting point to the process is also positive. Marlotte Herweijer, the process co-ordinator, emphasises, though, that each company must use the guidelines in its own way: 'You must make choices and not give any attention to the themes which are not important to your company or which you have no influence over.' She therefore recommended the following:

> Limit a questionnaire which is based on the OECD guidelines to the main points and take a good look at what you are already doing, because that is often quite a lot. Create a support base within the organisation by using the organisation's own language and by continuing developments which are currently taking place within the company. Give the subject attention at international meetings. Go with the flow: do not rush into things if the company is not ready for it. Do everything step by step and be careful not to start too over-ambitiously.

Corporate social responsibility themes in the policy	Examples of our interpretation
Transparency and responsibility	● The ambition to start reporting on corporate social responsibility based on the Global Reporting Initiative (GRI) ● To become a more transparent company (also concerning financial performance) and to enter into dialogue with the stakeholders
Chain responsibility	● To promote co-operation with the chain partners ● To promote chain management ● To observe standards and certification schemes
Governance	● To implement changes in the corporate governance structure and to also implement the Dutch Corporate Governance Code (Code Tabaksblat) ● To implement TQM (total quality management) principles
The environment	● To implement ISO 14001 ● To increase the market share of organic products
Employees	● To increase the involvement of the employees ● To formulate a company code ● To develop a model for improving the competence of managers
Human rights	● To accept the works council as a serious partner in discussions ● Some human rights issues are not considered to be as important, because Wessanen is mostly active in Europe and North America
Product responsibility	● To increase the strategic focus on authenticity and the traceability of the product to the source ● Increase food safety by developing our own guidelines ● To strive for healthier products

TABLE 2.1 Main corporate social responsibility themes: Wessanen

2.4 Conclusions

The plan of action explained above appears to be very useful in establishing some order in the maze of international guidelines and standards. The examples given by the companies all started by mapping out the current situation with regard to corporate social responsibility in an international context (see Action 1). The OECD guidelines for multinational enterprises are very helpful when doing this. In most cases, these guidelines were converted into a questionnaire tailored to the individual company. An informative meeting or another form of exchange of information is often organised before the questionnaire is sent, in order to explain the objective of the questionnaire. Local managers are sometimes a bit reluctant to answer the questions, because they regard it as a form of social control. There is also a fear of making things public, which may create legal liabilities or which may affect the stock price.

For some companies, the local branch managers are still unfamiliar with the corporate social responsibility theme and the OECD guidelines. In such cases, it is necessary to keep the questionnaire as simple as possible. The questionnaire is then sent to the relevant people within the worldwide organisation with a request to complete it. A few companies did not send the questionnaire round, but let the local managers or quality managers react to the questions by telephone. One of the companies did not produce a questionnaire, but instead used the guidelines as a topic of discussion for the management team.

Based on the questionnaire results, it is possible to tick off which corporate social responsibility themes still receive insufficient attention and, therefore, require more attention by the company. This assessment is often made during a meeting in which various company representatives participate. In this way, various competences and experiences available within the company are made use of. Remaining as close as possible to the organisation's daily practice appears to be of great importance when giving shape to corporate social responsibility, because this increases the support base within the company and leads to acceptance of the top-down approach, which is usually unavoidable when implementing this subject in the organisation.

It is the company's management team that then determines the policy priorities for corporate social responsibility. When deciding this, it is important to also include the expectations and demands of the most important stakeholders. Doing so will avoid the stakeholders questioning the choices made at a later stage. Theme-specific guidelines and standards can then be used to develop the selected priorities further. Maintaining a dialogue with stakeholders is also useful at this stage.

A code of conduct is the core of the company's policy with regard to corporate social responsibility, but it also seems to be a useful means to begin *internal communication* concerning the vision and the main features of corporate social responsibility in an international context. When drawing up a code of conduct, the OECD guidelines and some additional theme-specific international guidelines and standards are an important source of inspiration. It is a difficult and sometimes long-term task to allow all the people in the organisation to familiarise themselves with the contents of a code of conduct. It is crucial to use different types of communication, such as training programmes, workshops, written information, visual material and personal talks, to let the people know what the code contains, why it is important and what the company wishes to achieve with it. It is also important that top management visibly commits itself to the code of conduct.

A code of conduct also plays an important role in *external communication* of the company's policy on corporate social responsibility. The aspects that the implementation of the code of conduct relates to and the reasons for this must be explicitly indicated. Experience has taught us that all parts of the OECD guidelines and/or the theme-specific guidelines and standards are not equally relevant to each company. That is not a problem in itself, but it must be substantiated to the outside world, otherwise a company will be criticised because it does not give attention to certain matters. External stakeholders are often critical of company codes. Entering into dialogue with stakeholders concerning the contents at an early stage makes it possible to include their vision in the considerations and will also increase acceptance of the choices to be made.

The company should explain how it deals with discrepancies between local legislation and customs and international guidelines. Is the highest standard always observed or does this change from case to case? This matter will be dealt with in more detail in the next chapter.

3

Tension between observing international rules of conduct and local circumstances

3.1 Navigating between two extremes

The international guidelines and standards mentioned in Chapter 2 give companies a footing in the worldwide implementation of corporate social responsibility but, at the same time, they cause tension. In practice, it doesn't appear so simple to apply these standards in the same way across the board, as a result of the cultural and social differences between countries. Companies investing in a certain country must observe the national legislation and the host country's policies. Problems arise if these demands do not fully agree with or are less strict than international agreements. The question that international companies are confronted with is: to what extent should they conform to the local circumstances when implementing their own company policy?

In principle, companies must strive to have the same policy throughout the world and use the international agreements on corporate social responsibility as a basic principle for that policy. The use of universal values does not normally cause a problem where fundamental human rights are concerned, such as abolishing slavery, torture, forced labour and genocide. Dilemmas mainly arise in situations that are less straightforward. This is often the result of differences in cultural or political opinions between countries concerning specific subjects. A good example of this is the right to form free trade unions and collective bargaining for employees. In some countries, such as China, local legislation prevents this. What do you do as a company if your code of conduct states that universal human rights must be respected? If you, as a company, wish to recognise trade unions, but you are not able to do so, then you have a legal dilemma. Do you then decide to set up a type of works council as a compromise? The situation is also more complicated than it may seem at first sight where bribery and corruption are concerned. Practice has shown that, in some countries, it is difficult to organise something if you, as a company, are not prepared to pay some money. By paying very generous commissions to agents, companies can make themselves guilty of a veiled type of bribery.

In several cases, companies have found themselves faced with the dilemma of which position they should adopt. They can decide to go for a uniform company approach and observe universal principles or they can assume a cultural-relativistic approach and adapt to the local circumstances, a choice similar to balancing on a beam. If you, as a Western company, enforce your own values too strongly within branches located in a different culture, then you can be accused of cultural imperialism and remain an outcast from that society. On the other hand, if you do it the other way round and, as a company, totally conform to the local situation, then not acting in accordance with the international rules of conduct can be held against you by the international community (in particular, social organisations). The human rights policy of a country and trying to influence a national viewpoint are primarily government responsibilities. Social organisations share this opinion. They also recognise that states have a right to their own vision on themes such as human rights, which is also called the margin of appreciation. However, they consider this margin to be small.

In order to deal responsibly with the tension between international rules of conduct and local circumstances in concrete situations, many international companies are currently doing their best to draw up their own rules for their personnel. Instructions are drawn up and instruments are developed for each theme in order to make these dilemmas manageable. The following company examples will be discussed below: Shell's human rights policy, Heineken's integrity policy and Thermphos's environmental policy.

3.2 Human rights policy: Shell as an example

Shell is an international company active in energy and petrochemistry with branches in more than 145 countries. It employs a total of 115,000 people. Bert Fokkema, issue manager at Shell Nederland, emphasises that 'Shell considers it to be of utmost importance to deal with human rights issues properly. Core values within Shell are sincerity, integrity and respect for others. The human rights theme is closely related to this.' In 1997 Shell was the first company in its sector to include a commitment to human rights in its General Business Principles (www.shell.com/sgbp). These principles are applicable to all Shell subsidiaries, which means that they must be implemented in the company's daily practices. These days, Shell has an extensive policy concerning human rights. It now has:

1. A company code, business principles and management structures

2. Procedures (for example, the Environmental Social Health Impact Assessment [ESHIA])

3. An interactive training programme

4. Supporting documents for internal use to give country managers a number of practical guidelines on how to give shape to the human rights policy and how to deal with human rights dilemmas (for example, the management primers 'Business and Human Rights' and 'Human Rights Dilemmas: A Training Supplement', which are linked to a web training programme; see www.shell.com).

5. An annual report covering the three Ps (people, planet, profit)

The package itself is complete, but what is still missing is a method of assessment for finding out how well Shell deals with human rights and what risks the company faces if it doesn't observe them. Therefore, Shell used an instrument developed by the Danish Institute for Human Rights (www.humanrightsandbusiness.org).

3.2.1 Human rights assessment instrument

The Human Rights Compliance Assessment (HRCA) instrument consists of 1,000 indicators and more than 350 questions, based on the Universal Declaration of Human Rights, the core conventions of the ILO (International Labour Organisation) and other human rights treaties and conventions. They indicate the minimum responsibilities that companies have with regard to human rights, not only towards employees but also towards the chains and the societies in which they operate. The HRCA has been developed based on a wide consultation with companies and NGOs (non-governmental organisations) throughout the whole of Europe (www.hom.nl).

In view of the very many questions and indicators, the HRCA did not prove practical in this extended form. Therefore, together with the Danish Institute, Shell has developed a step-by-step plan to reach the desired result in a clear and concise way. There are three different steps:

1. Country Risk Assessment, CRA: drawn up by the Danish Institute. The objective is to identify each country's high-risk areas, with regard to its legislation and implementation. Based on this assessment, a number of focus areas are identified which companies and their chain partners must concentrate on, and a number of recommendations are also made. If a company is a member of the Danish Institute, it will have access to this assessment

2. Company assessment (Human Rights Compliance Assessment, HRCA): the company tests its own policy, procedures and practices against a number of selected questions and indicators

3. Strengths–weaknesses–opportunities–threats analysis (SWOT): the SWOT analysis enables companies to find out which opportunities they can make use of and which risks must receive attention. Important stakeholders can then be invited to discuss the plan of action resulting from the SWOT analysis

In addition to this step-by-step instrument plan, a short version called 'Quick Check' can also be used. This check selects a small proportion of the questions and indicators (28 questions and 230 indicators) from the total 350 questions and 1,000 indicators. The selection includes the indicators that are related to fundamental human rights and which are most applicable to companies. The advantage of this Quick Check is that it takes less time than the full HRCA and it can, therefore, be used to quickly gain an insight into the human rights situation. The disadvantage is that it doesn't include the full palette of human rights and it is not country-specific. The Quick Check can be found at https://hrca.humanrightsbusiness.org.

3.2.2 Results of the assessment instrument

In order to test the instrument described above, Shell had a country risk assessment (CRA) carried out in one country in the Middle East, one country in Asia and one country in North Africa. Shell also carried out its own company assessment (HRCA) in the first two countries. Since it is still in the experimental phase, the names of the countries will not be made public at this stage.[1]

The CRAs give the risk levels for each of the 20 universal human rights. From this analysis, transcending themes or *focus areas*, such as working conditions, are identified. These are areas to which a company must pay attention if it operates in that specific country. Table 3.1 shows the risk levels and focus areas for the three countries.

The results show that the Middle Eastern country has a reasonably well-developed regulatory system into which human rights are integrated. The five focus areas with recommendations for companies and their chain partners are: (1) salary, (2) forced labour, (3) privacy, (4) working conditions, and (5) religion.

The Asian country has very different scores. The formal legislation regulates human rights to a reasonable degree, but the legislation is not really applied in practice. This gives a high number of risk areas and a third score has, therefore, been added: namely, specific business risks. The focus areas that have been identified for companies and their chain partners are: (1) working conditions, (2) trade unions, (3) forced labour, (4) health, (5) salaries and terms of employment, (6) relocations, and (7) privacy and family life.

In the North African country, the business risk of violating human rights is generally a medium risk. Five focus areas have been identified for companies and their chain partners: (1) working conditions, (2) trade unions, (3) discrimination, (4) salary and (5) government relations.

---

1 In close co-operation with teams at Shell International and Shell Netherlands, Esther Schouten has carried out the test described here using the assessment instrument. She is a postgraduate student at the Erasmus University in Rotterdam, financed by TNO Quality of Life.

Score for the Middle Eastern country for the 20 human rights issues	High risk	Medium risk	Low risk	No. of focus areas
Formal legislation	3	8	9	
In practice	4	7	9	
Focus areas				5

Score for the Asian country for the 20 human rights issues	High risk	Medium risk	Low risk	No. of focus areas
Formal legislation	4	11	5	
In practice	19	1	0	
Business risk	13	7	0	
Focus areas				7

Score for the North African country for the 20 human rights issues	High risk	Medium risk	Low risk	No. of focus areas
Formal legislation	7	10	3	
In practice	11	7	2	
Business risk	5	11	4	
Focus areas				5

TABLE 3.1 Risk levels and focus areas for the three countries

An HRCA was then conducted in the Middle Eastern country and the Asian country. Forty-eight questions out of a total of 350 were selected for the Middle Eastern country using the country's risk profile and focus areas. The subsidiary in that country was also assessed in order to determine whether the necessary policy and procedures were in place to overcome the business risks. The SWOT analysis that followed showed that there were a number of strong points and a number of points that require attention. The strong points are that the subsidiary has strict and clear guidelines and control mechanisms with regard to security, as well as health and safety. Shell also provides training to help set up small local companies so they can eventually become Shell's business partners. Shell also provides water to outlying areas.

The examples in Boxes 3.1–3.3 show a number of points that require attention and the dilemmas encountered.

Forty-seven questions from the HRCA were selected for the Asian country. The local Shell company was also assessed in order to determine whether the necessary policy and procedures are in place to

The legislation in the Middle Eastern country forbids trade unions, although employers with more than 50 employees are obliged to set up a joint labour management committee to discuss work–related issues. However, the committee is not allowed to discuss salaries, working hours and employment conditions. Companies can, however, set up a staff committee in accordance with the law and slightly expand its mandate so that employment conditions can also be discussed. The HRCA recommends that not only national employees should sit on this committee, but also a number of the many low–skilled, foreign workers, so that they can also exercise influence regarding their employment conditions.

Box 3.1 The right to freedom of association versus national legislation

As a result of the laws in the Middle Eastern country, Muslims cannot marry non–Muslims and same-sex marriages are also not permitted. As a result of this, some couples live together, but are not officially registered as partners and their children are also not recognised. Therefore, the partners and children of such relationships have no right to social services. Furthermore, employees who have a religion other than Islam do not have the same rights as Muslims. Companies operating in this country must ensure that the employment conditions do not only apply to legally married partners and that non–Muslims can also enjoy flexible working hours, make use of prayer rooms and have paid leave for a pilgrimage to Mecca. The HRCA recommends checking the employment conditions for both Muslims and non–Muslims.

Box 3.2 The right to freedom of religion versus state religion

More than three-quarters of the working population in the Middle Eastern country's private sector consists of foreign employees, particularly from India and Pakistan. The government has, therefore, implemented a 'Nationalisation Programme' with the intention of reducing the number of foreign workers in the private sector by placing emphasis on the hiring of employees of the country's nationality. Companies are obliged to implement this programme, but implementing this programme can lead to discriminating measures against foreign workers. The HRCA, therefore, recommends companies to offer every (potential) employee the same opportunities and to give particular attention to foreign workers.

Box 3.3 The right to non-discrimination versus a programme of nationalisation overcome the business risks. The consequent SWOT analysis showed that this subsidiary has strict and clear guidelines and control mechanisms with regard to health and safety, as well as bribery and corruption. The company also showed that it is able to maintain a good relationship with the government. For example, the company is supervising a large relocation project as much as possible in accordance with international standards. However, Shell also encounters a number of dilemmas, examples of which are given in Boxes 3.4 and 3.5.

The Asian country has the death penalty. The death penalty is also given to people who have criticised the government or who have committed theft crimes. This results in a gross violation of the right to life. Furthermore, people are imprisoned without any form of legal process and they are very badly treated. Companies may find themselves confronted by this if, for example, employees or chain partners are imprisoned or if the government asks for private information about employees which can lead to imprisonment or the death penalty. The question is: how can companies deal with this? Practice has taught us that silent diplomacy and careful risk analysis provide better results than a public proclamation of your viewpoints. For every potential lawsuit, the legal service investigates whether this can lead to the death penalty and takes legal action based on the findings.

Box 3.4 The right to life versus the death penalty

Reports from human rights NGOs show that the government of the Asian country regularly monitors telephone conversations, fax messages, emails, text messages and internet communication. This leads to a violation of the right to privacy. Companies may also be faced by this and must, therefore, ensure that their information is protected. Visiting a website that is critical of the government can lead to the detention of employees. As an extra measure, companies can organise matters in such a way that international news can only be viewed by expatriates, because they cannot be tried according to the national law.

Box 3.5 The right to privacy versus government interference

3.2.3 Evaluation of the assessment instrument

Shell has found the use of the human rights instrument very valuable. The Vice President of External Relations, Policy and Social Responsibility for Shell International, Robin Aram, expressed it as follows:

> This tool looks promising because it helps to structurally identify sensitivities regarding human rights in our new and existing operations. Country assessments carried out by the Danish Institute give a quick and comprehensive overview of all the human rights issues in a country and pinpoint the top-of-the-mind issues for managers. Following this, company assessments help us focus our policies, procedures and performance on these sensitive issues, and provide useful input to dialogue with other stakeholders.

The results of the country assessments make it clear that the specific context of countries leads to many dilemmas for Shell's subsidiaries. The 'Management Primer', developed by Shell in 1998 concerning human rights (revised by Amnesty International and Pax Christi), states that many human rights are interpreted differently. This is because countries have different cultural values, which also change over time.

The main conclusion that can be drawn from the results is that it is possible to place consistent, minimum requirements on certain, existing, internal company processes throughout the world, although there are also aspects that have to be tackled locally. An example of this is freedom of religion. Islam is the national religion of the Middle Eastern country, while the government of the Asian country discourages any form of religion. This clear difference can, however, result in the same violation: namely, discrimination based on religion. Shell's basic principle in such cases is that the subsidiaries must have the company process set up in a way that does not discriminate against employees. How a subsidiary then solves a specific dilemma in practice (in this case, the type of religion) is left to the local management after consultation at corporate level. The local manager can then act within a certain bandwidth.

Besides indicating abstract, intrinsic frameworks (respecting human rights), a company such as Shell can also place clear demands on the internal process taking place locally. In other words, the local contents can be different (type of religion), but the process of solving dilemmas is the same. The HRCA instrument can be useful when determining this process.

Another point that became clear during the analysis of the current internal documents is that the emphasis is sometimes placed on observing national legislation, when international standards must also be observed. If national legislation conflicts with international standards (for example, if the national legislation forbids trade unions), it is useful to draw up a number of guidelines within the company.

The country assessments also show that the number of dilemmas increases the more the company infringes on territory where the government is the local regulating body. From the country assessment examples, it appears that working relations and discrimination, in particular, raise difficult questions. Relations with stakeholders can also produce conflicts, because engaging in dialogue with local stakeholders is often seen by the government as being a political activity. The Dutch *poldermodel*, a model where parties work together to come to a joint solution, can, therefore, not be used everywhere. Companies must deal with these relations carefully and try to operate within the frameworks of the country and the international standards. Quiet diplomacy is more helpful than taking a public stand.

Finally, it has been proved again that human rights are, politically, still a very sensitive subject. Human rights are often seen as standards that have been developed and enforced by the Western world. A way of overcoming this is, for example, to apply the human rights instrument in Western countries. The result could be that human rights are not necessarily perceived as Western imperialism by countries in the Middle East. Furthermore, it also appears that there are many differences of opinion concerning the ILO standards with regard to working hours, overtime, the amount of salary paid, social benefits and so on. The competitiveness of a company can be strongly influenced in emerging markets if all the ILO standards are strictly followed. The ILO has recognised this, which is why the ILO Fundamental Principles of Work were formulated in 1998 with four key principles: (a) no child labour, (b) the freedom to associate and organise, (c) no forced labour, and (d) no discrimination. Companies could start with these four principles.

3.3 Integrity policy: Heineken as an example

Heineken is active worldwide in the production, distribution and sale of beer, malt and soft drinks. Production takes place in approximately 120 production locations in almost 70 countries and the company employs approximately 60,000 people.

At the start of 2000, the Board of Commissions asked the corporate affairs department to draw up a code of conduct concerning integrity and a method of implementation for Heineken that had to meet the Dutch legal requirements and be relevant and workable in all the countries in which Heineken operates. A diverse project team was created for this task. In addition to corporate affairs, the department responsible for managing the project, corporate human resources and organisation development, corporate legal and business affairs, group internal audit, and corporate control and accounting were also represented.

The first question the project team was faced with was should the process be started top-down or bottom-up? The team chose bottom-up for various reasons. First, Heineken has a wide range of brands, 80% of which are local. Partly as a result of this, the relationship between the local subsidiary and the Heineken company is sometimes limited. Second, Heineken has to deal with many different cultures when implementing an integrity policy and it wanted to be sure that this was taken into consideration when drawing up the policy. Third, Heineken works according to a delegation model and the independent subsidiaries are largely autonomous. A top-down approach is, therefore, less in line with the company culture.

However, the subsidiaries are expected to work within a (broad) framework which is centrally determined. This particularly concerns the company values and basic principles, as well as the policy, which is formulated centrally. The translation of this central policy into a local policy is seen as a local responsibility. However, it is a more sensitive issue when it concerns the theme of integrity, because it is not so easy to have local deviations from this theme.

Several subsidiaries were involved in pilot studies when drawing up the integrity policy. It included the subsidiaries in sub-Saharan Africa, the Caribbean (Suriname), Italy, the United States and Indonesia. The following subjects were dealt with within the broad theme of integrity: legislation, anti-corruption, conflicts of interest, fraud, gift rules, chain responsibility and a whistle-blower code.

The following boundary conditions were taken into consideration when implementing the policy within the pilot subsidiaries:

- Each subsidary must make an analysis of the perception of risk for the subjects listed above. Next, it must indicate which positions within the organisation have to deal with these subjects to some degree and which of these positions are most vulnerable. Based on this, each subsidiary must draw up a local integrity code itself. When doing so, it must, however, meet the frameworks that have been drawn up centrally in accompanying documents concerning integrity. This forms the core of the implementation system

- Clear performance indicators and maintenance requirements must promote compliance with the policy

- The subsidiary must carry out the activities itself, but it is checked by the internal audit department (to make sure that the requirements are met) and by an external accountant (KPMG). Heineken has chosen *not* to go for an approach where individual employees must sign a declaration of moral conduct, because this is not in keeping with Heineken's control model, which is based on the delegation of responsibilities to the subsidiaries

- The implementation is not free from obligation. The local codes and the implementation plans will be tested by the Business Conduct Review Committee (a central testing committee). The way the implementation plans are carried out will primarily be tested by the internal audit group and an external accountancy firm

- A whistle-blower code has also been formulated. A whistle-blower has three contact options: a responsible line manager (on a confidential basis), a local confidential adviser or an external help-line, which communicates the issue anonymously to the integrity commission and the board of directors. The directors of the departments who also participate in the project team are part of the integrity committee

In order to be able to carry out the above, the following documents have been drawn up for support:

- A management primer (inspired by a similar Shell primer). This Code of Business Conduct Management Primer is a handbook for all the Heineken managers and describes which subjects fall under the integrity policy. It then explains which basic principle from its policy Heineken applies to each subject. The Primer also gives checklists and frameworks which will help to uncover an increased risk of fraud or corruption within the organisation and to find out under which conditions such action can take place. Finally, examples of dilemmas are given

- A Code of Business Conduct Implementation Manual. This manual has been drawn up to assist subsidiaries in the planning and implementation of the Code of Business Conduct activities. It explains step by step how they come to the desired implementation of the Code

● A Code of Business Conduct Whistle-Blower Code

The pilot studies showed clear differences in the way in which the local codes were drawn up. Some codes were very legal in nature, while others used many examples. Despite these differences, the testing of these local codes and the associated implementation plans by the Business Conduct Review Committee did not cause any major problems. The committee tested two main points: (1) Does the local code contain what it should contain? (2) Are the formulated basic principles reasonable for a certain culture (for example, is the limit for a certain gift reasonable)? Experience has taught Heineken that Heineken's central principles with regard to integrity (which are based on international guidelines) are a good starting point, but a certain local deviation must be possible in order to adequately deal with the diversity in cultures.

After the pilot studies had been successfully completed, the approach was extended to all the subsidiaries. Regional workshops were held in many countries throughout 2005 to discuss the Code. The Code was to be implemented everywhere by the end of 2005.

Heineken will do the following in order to ensure the Code is understood by all employees:

● All new employees will be informed of the Code and receive training

● Every two years, and also after every reorganisation, the subsidiary will be checked to see whether it still meets the Code

● Wide research among the subsidiaries concerning compliance with the Code will be carried out each year

● Works councils will be involved in the local implementation.

● Sanctions will be communicated

3.4 Environmental policy: Thermphos as an example

Thermphos is an international producer of phosphorus, phosphorus derivatives, phosphoric acid and phosphates, with production locations in the Netherlands, Germany, France, the United Kingdom, Argentina and China. The company employs a total of 1,200 people. Thermphos's management team strongly believes that corporate social responsibility is an essential prerequisite for the future success of the company.

The company, which was created in 1997 as a split-off from the Hoechst group, has seen rapid expansion over the last few years, partly through acquisition. As a result, the need emerged for more corporate guidance from the head office in Vlissingen, the Netherlands, with regard to corporate social responsibility.

Environment is the single most important theme for Thermphos with regard to corporate social responsibility. Hence management decided to start streamlining the environmental policy.[2] The starting point was compliance with the OECD environmental guidelines, combined with certification of all sites according to the ISO 14001 standard. As a zero baseline, a compliance check using a questionnaire was carried out. The check raised some questions with respect to differences in culture and circumstances when implementing an environmental policy. A study was conducted to provide an answer to that question. The study initially focused on the three most diverse parts of the company: namely, the Thermphos companies in the Netherlands, Argentina and China. The companies vary in size: the Dutch branch has approximately 450 employees, the Argentinean branch has approximately 70 employees and the Chinese branch has approximately 250 employees. The Chinese company is a joint venture with 60% of the shares owned by Thermphos and the remainder by Chi-

---

2   Management was assisted by Eddy van Hemelrijck, who carries out postgraduate research at the Copernicus Institute at the University of Utrecht.

nese partners. Thermphos in the Netherlands and Thermphos Argentina are 100% subsidiaries of Thermphos International.

First, Thermphos wanted to obtain an impression of the state of affairs within the three branches with regard to the environmental policy. Discussions with the management of the subsidiaries highlighted some clear differences in the degree to which they thought the local company's environmental policy agreed with the OECD guidelines. Argentina carried out every sub-aspect of the guidelines well to very well as far as the environment was concerned. The Netherlands also scored well, but with regard to co-operation with external parties (customers, government institutions, people living close by, etc.), the need for a more structured, proactive policy was recognised. In contrast, China clearly scored lower with regard to the OECD guidelines concerning the environment, although they did meet the local environmental legislation. As proof of this, the Safety and Environmental Protection Award issued by the Xuzhou Environmental Authority in 1999 and presented to the company by the local government was shown.

Employees were also asked for their opinion on the environmental policy and environmental management of their respective companies. Interviews were held with 10% of the employees of the Dutch branch, 16% of the Chinese branch and 25% of the Argentinean branch. In the Netherlands and Argentina, 75% of the employees interviewed were satisfied with the environmental policy and did not feel the need for additional measures. On the other hand, in China 88% of those interviewed indicated a need for the company to invest more in its environmental management. The opinion of the employees alone makes it obvious that the same environmental measures must be implemented in the Chinese branch as in the Netherlands and Argentina.

The second part of the study aimed to provide an answer to the question of the extent to which these differences are related to differences in the local context and/or the internal management of the company. In order to answer this question, the local differences between the three countries (in particular, the relation between government and business) and the differences in the organisational structure and leadership style were examined.

Based on the Environmental Sustainability Index (ESI)[3] (2005), the Netherlands scored the highest (eighth place) of the three countries that were examined with regard to the 'social and institutional capacity to react to environmental challenges' (this includes the government's environmental policy, eco-efficiency, the ability of business, science and technology to respond). Argentina came 31st and China 88th. Additionally, in order to obtain an image of the appreciation for the government's environmental policy, a small, random sample survey was held among future managers.[4] It involved written interviews with 208 Argentinean, 62 Dutch and 83 Chinese business management students. Of these students, 94% of the Argentineans, 58% of the Chinese and 36% of the Dutch were of the opinion that their government did not do enough to protect the environment.

The information given above shows that the governments in Argentina and China have less of a regulatory role than the government in the Netherlands. The fact that the Argentinean branch rates itself so positively and the Chinese branch rates itself so negatively with regard to the environmental policy has, therefore, probably more to do with internal factors rather than local factors. Interviews with the management and employees of the three locations confirm this conclusion.

Management changes have taken place over time within the three branches. The most stable is Thermphos Argentina, which has had the same Argentinean General Manager for more than ten years. The Chinese subsidiary has had a number of managers with different nationalities within a short period of time. The Dutch company was initially managed by a German native, who was subsequently succeeded by two different Dutch men and an Argentinean from 1999.

---

3   YALE, www.yale.edu/esi/ESI2005.pdf consulted on 29 January 2005.
4   This work was carried out by Eddy van Hemelrijck, postgraduate student at the Copernicus Institute, the University of Utrecht.

The environmental policy is organisationally embedded within each branch in a different way. The person responsible for the environment in China reports to the production manager. In the Chinese culture (with large distances between two levels of power), this means that the person responsible for the environment has little or no power and, definitely, no standing. In the Dutch and Argentinean branches, the person responsible for the environment has a staff function. In the Netherlands, he reports directly to a member of the board of directors and, in Argentina, directly to the general manager.

From the interviews with employees of the three branches, it appears that the general manager of the Argentinean organisation scores exceptionally highly with regard to care for the environment. In view of his long-term service as the general manager, he has been able to put his mark on the organisation and the company culture. This explains why the Argentinean branch, despite a poor government policy, performs very well with regard to the environment. On the other hand, the Chinese management has constantly changed and, partly as a result of this, a consistent environmental policy has not been developed. Also, the pressure from the government to do so has been limited. Finally, the Netherlands, partly under the influence of the government and other pressure groups, has had a stringent environmental policy throughout the years.

Differences in the environmental policy within the three assessed branches of Thermphos therefore appear to be predominantly determined by internal factors and more specifically the leadership of the local management and less so by the local context (i.e. local appreciation of environment). The observed differences in environmental performance are, in principle, bridgeable because many of the environmental measures to be taken are of a technical nature. Based on this observation, the management of Thermphos has decided to employ one generic environmental policy within the entire company. The company has the following arguments in favour of this:

- A generic environmental policy forms an extra binding agent in the organisation

- Thermphos wishes to be proactive with regard to the environment

- Thermphos wants to be able to deliver to its international customers from any branch. To do this, it must be able to guarantee these international customers that the policy and the quality that result from this policy are the same, regardless of where the product has been produced

The implementation of one generic environmental policy requires good communication between the central level and the local branches. The local manager and the employees must be able to translate the guidelines developed at central level with regard to the environment into their daily practices. From the interviews with employees from the three branches, it appears that the most effective method of communication depends strongly on the specific local culture.

Argentina is characterised by a high degree of collectivity, while the Dutch are much more individualistic and, in China, the hierarchy of power plays a central role. When asked how the new policy rules could best be communicated, the employees gave varying answers in line with the differences stated above. In Argentina, the preference was for strong, verbal communication to the group, combined with written material in as many different places as possible (company brochure, company panels, etc.). In the Netherlands, a large degree of participation was expected and desired, as well as a very direct, individual approach. In China, they wanted only clear, preferably written, instructions.

In order to guarantee success, the employees in both the Netherlands and Argentina underlined the importance of the example set by the management. The importance of the environmental regulations also had to be translated into reporting systems, performance premiums and so on. Both countries also requested the implementation of concrete environmental projects and Argentina requested to do this in small, multifunctional groups. The Chinese, however, expected strict supervision and sanctions if the rules were not observed. The Chinese employees who were interviewed did not believe that the policy would be implemented if there were no sanction policy, regardless of how efficiently it is communicated.

From the Thermphos example above, we can conclude that a company can, in principle, implement its environmental policy within all of its subsidiaries in the same way. In contrast to a policy on human rights and integrity, ethical and social considerations hardly play a role, because an environmental policy is of a more technical nature. The *what* can, therefore, be determined at a more central level. The challenge lies more in the way in which the central level leads and communicates and how the local management picks up on the theme and implements it in the organisation. The *how* must be interpreted locally. This must be taken into consideration when rolling out the generic environmental policy to the first layer of managers.

3.5 Conclusions

From the experiences gained at Shell, Heineken and Thermphos, it appears that the social themes, such as human rights and integrity, create more moral, culture-related dilemmas than the environmental theme. Problems arise for subjects such as free trade unions, discrimination and equal opportunities (men/women, religion, sexual orientation), work times, the protection of privacy and the rules concerning gifts. Moral and culture-related considerations play an important role in these social themes and it is therefore more necessary to take the local customs and circumstances into consideration. It is not possible to outline in detail from a central office which conditions local branches must meet for various social themes.

In order to guarantee uniformity in the policy implementation, both Shell and Heineken have drawn up some rules. Both companies place demands on the internal, local process. The local managers must indicate which dilemmas they are confronted with and how they wish to solve them. Through consultation with people at central level, it is then determined whether the proposed solutions are in keeping with the general policy. A bandwidth must be determined for each item to indicate within which limits the local management may act. This bandwidth must be in line with the policy that has been determined at the company level and it must also be in keeping with international guidelines. By maintaining clear and unequivocal procedures, both companies have found a way of dealing with the field of tension between observing international rules of conduct and local circumstances.

In order to give external stakeholders a transparent image of the rules regarding the social themes that a company such as Shell or Heineken observes, it is not sufficient only to communicate the code of conduct. Since a code of conduct contains more general phrasing than is used in practice, a more complete image is obtained by giving insight into the procedures and regulations that the company has drawn up in order to support the internal decision-making process. Even then, it will still be difficult for external parties to assess a company's actual behaviour, because the pros and cons are considered per item and are dependent on the local context. In order to increase the possibility of external assessment, insight can be given into the bandwidth within which the company branches must act for a certain theme and into the reasons why this bandwidth has been chosen.

When developing the environmental policy, a company can, in principle, set the same performance requirements worldwide. This does not mean that ethical dilemmas are not an issue, as is proven by the discussions concerning the use of genetically modified organisms. The choices made, though, are considerations by the management of a company rather than choices motivated by cultural differences. However, it is possible that standardisation of a policy is, in practice, limited by the large differences in the level of technology within the various production locations. Climatic and soil differences and the presence or absence of certain natural resources may make regional differentiation necessary (for example, with regard to the use of water, sources of energy and fertilisation).

Whether a company is actually able to implement uniform standards with regard to the environmental policy largely depends on internal factors. Based on Thermphos's experiences, it appears that the vision of the local manager and his personal leadership style are an important key to success. The position that the environmental policy occupies within the organisational structure (in particular, the authority of people responsible for it) is crucial. Finally, it is important to keep the communication style of the company's management in line with the culture of the branch. The most effective mix of communication methods differs between countries.

The above-mentioned conclusions based on the experiences of Thermphos do not apply only to the implementation of an environmental policy. The management style and the method of internal communication also partly determine the success of implementation of other policies. It is, therefore, important to pay special attention to this, also when implementing a social policy.

Experiences with using international guidelines and standards (Chapter 2) and the use of these in local situations (this chapter) lead to the plan of action shown in Box 3.6.

1. Assess the current situation with regard to corporate social responsibility within the company and its branches and any relevant chain partners based on the OECD guidelines for multinational companies

2. Formulate a preliminary vision and code of conduct regarding corporate social responsibility based on the results of step 1

3. Enter into dialogue internally and externally with relevant local and international stakeholders about their expectations and demands and reformulate the preliminary vision and code of conduct based on the findings

4. Develop short-term and longer-term strategies for corporate social responsibility in an international context and use them to draft a plan of action. Make use of theme-specific guidelines and standards

5. Integrate the approach chosen as much as possible into existing company processes and procedures. Set up a monitoring and reporting system. Make use of indicators to assess the progress made. Take into account the location-specific interpretation of corporate social responsibility within a centrally determined bandwidth. Draw up specific procedures and support documentation for this

6. Embed the process by incorporating the chosen approach into the quality and management systems. Assign responsibilities and tasks to departments and personnel and integrate these into the existing remuneration systems. Turn the approach into a continuous process of improvement

7. Communicate the approach and the results obtained internally and externally. Adapt the style of communication to the culture of the company branch and the local, external stakeholders. Ensure that there is visible commitment from senior management

Box 3.6 The plan of action for implementing corporate social responsibility in an international context

4

Corporate social responsibility in different political cultures

4.1 Country-specific characteristics

Practice has taught us that companies developing their policy with regard to corporate social responsibility must not only take the socio-cultural context into consideration, as discussed in Chapter 3. They also need to pay attention to the social needs and problems that exist in a certain country. For example, black empowerment and contributing to the fight against AIDS are central themes in South Africa. In the United States, corporate social responsibility is often associated with charity and, in the Netherlands, it is seen as activities that transcend legislation. The political–social situation in a country plays a large role in what is expected from companies—and certainly from foreign, Western companies—with regard to corporate social responsibility. Something that is obvious in one country can be a very important topic of discussion in another. This is the result of differences in:

- The social problems that are given priority in a certain country

- The relationship between (multinational) companies and the local government

- The relationship between (multinational) companies and their stakeholders (including social organisations) and the role of the citizens

How these differences can affect the approach to corporate social responsibility will be illustrated below by some examples.

First, two examples will show how each country gives its own meaning to corporate social responsibility, depending on the urgency of certain social problems and the specific political and sociocultural context. The two countries are China and Brazil.[1] Both countries are rapidly developing economies, but the priorities companies have with regard to matters of corporate social responsibility are very different. It is important for companies to know this before they invest or enter into any business relationships in that particular country.

Next, the experiences of three companies, operating in different countries, will be discussed, namely:

- ABN AMRO, whose home markets are the Netherlands, Brazil and the United States

- Pentascope, which was involved in the privatisation of a telecommunications company in Nigeria

- Koninklijke Houthandel Wijma (Royal Wijma timber trading company), which has been logging in Cameroon for decades

---

1 The information about corporate social responsibility in China and Brazil is based on a literature study and two brainstorming meetings with experts, which were organised within the framework of the 'Corporate Social Responsibility in an International Context' programme.

4.2 Country-specific interpretation of corporate social responsibility: China and Brazil as examples

The meaning of corporate social responsibility in a specific country is determined by the social priorities and the political and sociocultural context. The following two examples will illustrate this.

4.2.1 China

China had a tradition where, from a social point of view, corporate social responsibility was strived for before it was even known what it meant. The Communist state provided all the necessary facilities, from the cradle to the grave. Since the emergence of state capitalism, these social facilities have mostly been dismantled. In those days, the environment hardly played any role.

If people in China now talk of corporate social responsibility, then they are usually referring to partial aspects of it. With regard to the environment, for example, it is associated with following legislation. The gap between strict national legislation and the observance of it at a local level is very large. Therefore, merely observing legislation is considered quite something in itself. Fast economic growth and the absence of middle management to keep growth on the right track are high on the political agenda in China. There is a high level of unemployment among the unskilled workforce (approximately 300–400 million). Many of these unemployed have moved from the countryside to search for work in the cities and do not have an official place of residence. This group of employees is treated very poorly. They are paid very little and live in terrible conditions. Since they are often employed for short periods of time, they lead a nomadic existence. Therefore, good food supplies and a place to sleep are basic needs for Chinese employees. The next issue is the basic minimum pay. Philanthropy plays no role in China.

Chinese companies who operate internationally are more suscepti-ble to corporate social responsibility than other local companies. They are more often confronted with pressure from purchasers from, in particular, Europe and the United States to meet international codes of conduct and they are regularly audited by third parties for this purpose.

Western companies that have invested in China are usually front-runners with regard to the environment and social issues. In China, it is not unusual for workers to work in Chinese factories for a pittance and sometimes not to be paid at all or far too late. Chinese factories often also have very environmentally unfriendly production pro-cesses. For international companies, the fact that China is still a low-wage country is very attractive from an economic point of view, but to take advantage of this is bad for their reputation and, therefore, not in their interest. When international companies start co-operating with Chinese partners (for example, in the form of a joint venture), there are often conflicts over the strict environmental and social demands that the international companies want to place on their Chinese part-ner. The Chinese partner considers this to be a hindrance to rapid growth. Experience has shown that scepticism from the Chinese part-ner can often be removed by implementing concrete environmental and social projects together and, if these projects are aimed at cost sav-ings, the partner will often quickly see that corporate social responsi-bility can be useful. Also in the event of a takeover of a Chinese com-pany, there are often problems concerning the environmental and social legacy that the company leaves behind. International compa-nies must, therefore, be aware of the expectations of the local institu-tions and governments concerning a company's performance with regard to environmental and social issues.

When a multinational company makes a large investment in China, the government expects this company to use as many Chinese suppli-ers as possible. Since the Chinese government assumes that this com-pany uses modern technology, the company can place demands on the quality of the suppliers and it can ask Chinese companies to meet certain standards. This selection of good-quality companies has a stimulating effect within the Chinese business sector to take corporate social responsibility seriously.

The most important stakeholder for the international business sector is the local government. If an international company wishes to run a successful operation, it is a good idea to find a person or organisation in the region who knows all the ins and outs of the local government, otherwise it will be very difficult for the company to find its way. Employees and, in particular, middle management are also important stakeholders. Although there are NGOs in China, they are usually an extension of the government because, without sponsorship from the government, an NGO can do little or nothing in China. There has been a rapid increase in the number of environmental NGOs. Their most important task is information and education, but the number of legal actions against companies that pollute heavily is also growing. An overview of the social organisations present in China can be found in the report 'Corporate Social Responsibility in China: Mapping the Environment' (see www.theglobalalliance.org).

In January 2005, the Chinese Business Council for Sustainable Development was established as a sub-organisation of the World Business Council for Sustainable Development. Chinese and foreign companies wishing to get involved in the field of corporate social responsibility sit together in this Chinese Council.

4.2.2 Brazil

In Brazil, corporate social responsibility is particularly associated with social commitment. The large contrast between rich and poor and the discrimination against minorities in the labour market lead to a number of specific priorities. Companies are expected to give a high priority to, for example, an acceptable relationship between the highest and lowest salaries, the number of black employees and their average income, the average income for women versus that of men and the question of rightful ownership of ground, but the subject of health and safety in the workplace also deserves a great deal of attention. Being a good employer is perceived as being the basis of corporate social responsibility. If you, as a company, contribute to society on top of that, then you are really good. Philanthropy is, therefore, also an important spearhead. This could, for example, be making a financial or personal contribution to education and the healthcare service or participation in government commissions and other government bodies involved with social issues. For example, a contribution is expected from business to President Luiz Inácio Lula da Silva's 'No more hunger' campaign. Such national initiatives often enjoy wide support. Corruption, the payment of taxation to the government and crime are also points of attention.

The environment plays a much less important role and it is usually perceived as being a problem caused by Western countries. This has now led to measures being taken which also concern countries such as Brazil. That is, at least, the general feeling in Brazil. People have the feeling that the West has limited Brazil's possibilities for using its own resources for economic growth.

Brazilian culture is rather nationalistic and proud and this leads to double standards. If a Western company does something wrong in Brazil, then it is front-page news. If, on the other hand, a Brazilian company does something wrong, it may not even be mentioned in the newspapers. Western companies that are less well known by Brazilian citizens or which have local management have less to fear from such damage to their image. It is noticeable that a group of companies joined together at the end of the 1990s to form Ethos. This organisation is linked to Business for Social Responsibility and its ambition is to promote corporate social responsibility within the Brazilian business sector. One of the things it has done to achieve this aim is create its own benchmark in order to compare the social performance of companies.

The most important stakeholder for a business wishing to invest in Brazil is the government. Social and environmental legislation is well developed and the enforcement system also works reasonably well, particularly in the south of the country. Furthermore, the many independent trade unions (such as Observatorio Social) have a great deal of influence. Less important, but still influential, are the NGOs, which know how to increase their power via alliances with NGOs from Western countries. There is not much pressure from individual consumers and customers for companies to accept their corporate social responsibility. Brazilian NGOs tend to have a broader interpretation of the corporate social responsibility theme than their colleagues in Western countries. They associate it with a culture of responsibility that is broader than the company. They connect it with fundamental symptoms from local society, such as discrimination, inequality, corruption and a lack of democracy.

4.3 Corporate social responsibility in different countries: ABN AMRO, Pentascope and Koninklijke Houthandel Wijma as examples

Companies that operate in an international context all have to deal with differences in the way in which countries interpret corporate social responsibility. Three examples are given below: ABN AMRO, Pentascope and Koninklijke Houthandel Wijma.

4.3.1 Example: ABN AMRO in the Netherlands, Brazil and the United States

ABN AMRO is a bank with approximately 98,000 employees. The history of the bank goes back to 1824 and it now has branches in 58 countries. Its home markets are the Netherlands, Brazil and the United States. The local cultures of these three markets are very different and those differences can be seen in the interpretation of corporate social responsibility by the local bank.

It is important to know that the bank has developed a generally applicable policy with regard to corporate social responsibility which can be specified by each market. This general policy contains uniform company values and principles which every employee must take into consideration. These basic principles are subscribed to and propagated by the top of the organisation and they are a part of the company's core processes such as granting credit, how to approach customers and employees and the bank's social commitment. That this last point can stretch further than the immediate local environment may be obvious from the fact that the Millennium Development Goals of the United Nations serve as a guideline for the contents of core activities. Examples are microfinancing in Brazil and India and paying attention to reducing the use of fossil fuels and CO_2 emissions (for example, by reducing the company's energy consumption and by including criteria in the granting of credit).

The Netherlands

The Netherlands is a prosperous country, which was confronted in the early 2000s with an economic recession that has affected employment opportunities. The number of jobs at ABN AMRO was reduced by approximately 7,000 in 2000/2001. Political developments at this time also led to an escalation of the political debate concerning, for example, the multicultural society, and a number of scandals in the business sector have also damaged public trust.

It goes without saying that ABN AMRO has been affected by these developments. It is increasingly being called to account by consumers (who vote with their feet) and by NGOs. Openness and transparency are considered necessary, while the bank is also expected to be prepared to listen to criticism from society.

ABN AMRO recognises this concern and considers it very important to react to this as openly and as constructively as possible. The bank does this by attempting to engage in dialogue with interest groups, which range from the Friends of the Earth Netherlands to Amnesty International. Questions of a social, ethical and environmental nature are also increasingly becoming part of customer acceptance, risk assessment and financing. ABN AMRO attempts to be as transparent as possible with regard to who and what the bank finances and what the consequences of this are. Examples are the financing of defence equipment and environmental activities, such as felling trees and the extraction of oil and gas. A sustainable purchasing policy is also being developed (see Chapter 5).

The bank also actively contributes to combating social problems that have a logical connection to its business activities. An example of this is that it contributes to making the financial sector more accessible to migrant entrepreneurs while, at the same time, offering long-term coaching. The bank also spends time on being a good employer, which means that, besides a good salary, other important issues are education, growth opportunities and diversity, regardless of the employee's age. The bank is considering including in its leadership criteria the stipulation that visibly giving meaning to corporate social responsibility is a condition for promotion.

Finally, the bank considers it important to make its own contribution to the environment, which is why the reduction in energy consumption and CO_2 emissions are two priorities.

Brazil

The social situation in Brazil is very different from that in the Netherlands. The country has large differences in wealth within its own population and has considerable cultural diversity. The problem of poverty is high on the political agenda, and certainly since Lula became president in 2002. Prior to becoming president, he was a metalworker and trade union leader and he clearly sees himself as a representative of the lower classes. Social commitment plays a key role in the decision-making process within Banco Real (the Brazilian bank of ABN AMRO). The bank's management and the director, in particular, think the bank should also make the social problems in the country its own responsibility, obviously within the bank's sphere of influence. This fundamental social attitude is actively communicated within the bank's training and education programme (the Training Academy). Banco Real also provides insight into financial administration and the bank issues microfinancing to entrepreneurs in slum areas. Emphasis is also placed on a policy of diversity. Finally, priority is given to the environmental consequences of the activities to which the bank extends credit. This final point gained attention after a Brazilian NGO had criticised the bank regarding this point. The management took this criticism seriously and asked this organisation to give all of the bank's account managers a two-day training course on the environmental knowledge needed to grant credit. This was definitely an exception in a country where a culture of consultation is not the norm (the contrary is rather the case) as it is in the Netherlands. The training did indeed take place.

The United States

In the United States, large companies are expected to play an active role as leaders of society. For example, the representatives of companies often sit on committees in their free time. Social organisations try to attract good administrators from the business sector in order to get enough connections with politics and the business world. As a result, a network of parties with mutual interests is created. For example, in Greater Chicago, all those people with some kind of social role know each other. Many informal consultations take place within this structure and a good name in this community is essential for the company to function well.

LaSalle, the US bank of ABN AMRO, is also socially active in the way described above. Approximately 400 employees are on the committees of approximately 700 social organisations, varying from large museums to organisations involved in fighting poverty by redeveloping run-down neighbourhoods. At the same time, LaSalle fulfils a pioneering role in areas linked to the bank's activities. An example of this is the conscious choice to be the largest business sponsor of the Tax Assistance Program. More than 800 tax return forms have been completed by 138 volunteers from the bank for citizens with a below-average income, taking advantage of every possibility for financial relief which would have otherwise remained unused as a result of ignorance.

The government finances socially relevant developments or institutions to a smaller extent than in the Netherlands. That is an important task for the private sector, which has a number of consequences for LaSalle:

- In view of the country's political structure, LaSalle must make donations to both political parties—in reality, there is no opt-out

- Charity is considered to be a moral obligation and is, therefore, no longer voluntary. Donations from companies are essential for society to operate. This applies to, for example, public radio, museums and social work. LaSalle provides financial support to 700 NGOs

- Items that are not on the political agenda are, therefore, less important from a company's perspective. A vision on climate change and on the Kyoto agreement that differs from Europe's is an example of this

The government attempts to compel private financing in various fields through legislation. An example of this is the Community Reinvestment Act, which states that banks must also make a contribution to the redevelopment of deprived neighbourhoods. LaSalle is actively involved in this.

ABN AMRO, therefore, places different emphasis on the development of a policy with regard to corporate social responsibility in its three home markets, which is also expected of them by these markets.

4.3.2 Example: Pentascope's involvement in the privatisation of a telecommunications company in Nigeria

Pentascope is a consultancy that implements changes in organisations. As a continuation of this core activity, the company sets ambitious goals concerning corporate social responsibility, particularly with regard to its internal social policy. Pentascope was established in 1990 and has approximately 300 employees. In the early 2000s a project was initiated in Nigeria to aid the privatisation of the state telecommunications company Nitel. This project was a tremendous challenge for Pentascope. The privatisation of Nitel required a large, internal cultural change. Previously, customer relations were mainly determined by family ties. For example, if you were a cousin of one of the employees and you had problems with your telephone, then Nitel sorted out the problem, often free of charge. The question was whether and how Pentascope, as a Western company, could bring about changes in this local culture.

A contract was signed with Nigeria at the start of 2003. Pentascope first took a couple of months to investigate the situation in Nigeria and then drew up a business plan which included a number of components:

- Bring the telecommunications network up to standard

- Implement improvements in the organisation

- After achieving this, expand the network and offer more services

The financial administration also had to be improved. The objective was to have 90% of the invoices paid.

Initially, everything looked good. The head of the privatisation committee and the responsible minister both supported the privatisation and positively supported Pentascope's role in it. However, the situation changed when elections were held in April 2003. The president was re-elected, but the head of the privatisation committee became a minister and his original position was taken over by somebody else who had a more negative attitude towards the project. The new telecommunications minister also had a different view from that of the previous minister. Pentascope's most important allies were replaced by opponents.

It was in the interest of the new minister and also the new head of the privatisation committee to stop the privatisation. In Nigeria, ministers are chosen with the support of chieftains. In order to maintain their positions, they are obliged to do favours for these chieftains after the election and they are expected to act in accordance with the wishes of these chieftains. They cannot pay their debts from their regular salary and this must, therefore, be arranged in another way. By privatising Nitel, the new minister would lose an important source of income. Therefore, the new minister quickly let it be known that he wished to revise the contract with Pentascope. Other parties then also started to criticise the privatisation of Nitel, either directly or via the media and the negative mood this created did not make Pentascope's work any easier.

Investments had to be made in order to be able to implement the privatisation plans for Nitel and money had to be borrowed for that. However, the banks wanted a guarantee from the government that the loans would be recognised, but the government was not prepared to provide this initially. This created a huge delay and the original investment plan of approximately US$1 billion could not be implemented. Initially, everybody wanted to be able to have influence over this plan in the hope of getting a piece of the pie themselves. As soon as they were granted a contract, their criticism subsided, but Pentascope did not want to work this way.

Pentascope had previously agreed a fixed fee and a bonus with the Nigerian government. There was hardly any room for manoeuvre in the fixed fee (approximately US$10 million) and any earnings would have to be made from the bonus. When things became difficult, even the fixed fee was not enough for the project to continue. The fixed fee–bonus ratio was non-negotiable. Another problem was that the value of the companies to be sold was kept as low as possible by the investors, because they had an interest in using the press to do so.

Finally, Pentascope took the plunge and withdrew from the project. Pentascope wanted to request an independent assessment from an arbitration committee, but the Nigerian government foiled that idea as well. It called on an obscure law which stated that Pentascope was not registered as a company in Nigeria and, as a result, the contract was declared to be unlawful. The result of this was that arbitration was not possible. Pentascope then ensured that its personnel and financial damage was kept as small as possible.

Josbert Kester of Pentascope, who was closely involved in the project, learned a number of lessons from the experiences gained in Nigeria:

> You need a joint venture with a party close to the people in power. It is risky to invest there as a company on your own. You must have a sponsor at the highest level. This is essential for the success of the undertaking. A different approach is to pay bribes or grant contracts. Nitel's competitor, Globacom, has, for example, given 20% of its shares to the vice president for free. This is a common method, but is not Pentascope's way of doing business. In Nigeria, people talk very openly and without embarrassment about this type of corruption. The irony of it is that if you do not participate in such activities, you are accused of corruption yourself! This way of doing business could no longer be reconciled with our own company philosophy.

4.3.3 Example: timber extraction by Koninklijke Houthandel Wijma in Cameroon

Royal Timber Trader G. Wijma & Zonen BV (Koninklijke Houthandel Wijma) is a timber company established in 1897. The company has a long tradition in managing the entire chain from timber harvest in the forest, via timber processing and trade to the final user.

Approximately 70% (volume) of the wood traded by Wijma comes from its own production facilities (sawmills). The other 30% is purchased from third parties, particularly from West Africa. Wijma has been active in this region for many years. As a result of redevelopment after the Second World War, there was a large demand for hardwood in Europe. Azobé timber was introduced by the Dutch Ministry of Transport, Public Works and Water Management as a good alternative to oak. As early as 1968, the supply of round timber from West Africa was guaranteed by establishing a purchase office in Cameroon. Today, this company has grown to become Wijma Douala S.A.R.L.

Since the 1990s, Wijma has been involved in a conflict between the timber sector and various environmental organisations. At the time, these organisations called for a boycott on the use of tropical timber. Some time later, they slightly changed their position and the emphasis was placed on promoting certified sustainable forest management based on the principles and criteria of the FSC (Forest Stewardship Council). It was difficult for Wijma to obtain this certification immediately. Mark Diepstraten, the company's environmental co-ordinator, explains why:

> Wijma has the misfortune of having been active for many years in the trade of timber from a region where there are no or hardly any certified forests. For various reasons, certification, or demonstrable sustainable forestry, often only comes off the ground very slowly. Firstly, many countries in the Southern hemisphere lack the experience, knowledge and financial resources found in Western countries. Secondly, developing countries find it difficult to form an image of the demands which Western countries wish to place on them. This is caused by the diversity of opinions the various stakeholders have, such as the consumers (and/or the environmental movement), science, the politics of international donors and the policy of the various countries. The requirements to be met from the very beginning are also often too high.

In order to find a solution to the problems given above, Wijma has chosen to expand its volume of certified wood in phases. To do this, the company had to deal with many different problems. Large-scale, long-term forest concessions were necessary in order to make certification possible. The company initially obtained these in 1996 from the government in Cameroon, but they had to hand them back in again as a result of a reform in the forestry legislation and the linked land-use planning. This meant that Wijma was, once again, dependent on small-scale (non-sustainable) forestry exploitation for its raw materials. During this difficult period, environmental organisations raised the problems of illegal timber and started promoting the use of FSC-certified wood. Since Wijma's defence was not transparent enough in the eyes of the critics, the company appeared to be a willing victim.

Since 2001, Wijma once again has an area of long-term concessions in Cameroon (currently a total of 190,000 hectares for the next 30 years). These forest areas have been designated as production forests by the government and are, as such, allocated in the land-use planning. A forest manager is obliged to obtain a management plan for the forest area which has been approved by the Cameroon Ministry of the Environment and Forestry within three years after being granted a concession. This management plan must state how the forest manager will manage and harvest the forest in a sustainable way, taking into consideration the sociocultural, economic and ecological aspects of the forest. The forest is divided into 30 sections in the management plan. One section of the forest is harvested each year where only two to three trees per hectare may be felled. It is not until 30 years later that a particular section will be harvested again and the forest will have naturally recovered during that time. By the end of 2005 Wijma had completed the certification process for its first concession.

The process of trying to obtain FSC certification in Africa has literally been pioneering work. Mark Diepstraten explains the lesson learned by Wijma from the above experiences as follows:

> We should not turn down social discussions, but participate in them. We must communicate our chosen path clearly and transparently. You must, as a company, communicate what you do, why you do it and how you are going to achieve it. The internal co-ordination and communication within Wijma was not optimal at first. Lately,

however, things have changed. The new director in
Cameroon is more open, because he understands how
important it is to make a link between the market requirements in Europe and timber extraction in Cameroon.

The visions within Wijma are starting to come together. Wijma's
mission is to be a reliable supplier of a reliable product. Besides issues
such as quality and delivery time, legality and sustainability are also
consistent with this mission. Wijma is committed to achieving a level
of best practice for all its activities; in some countries this can turn out
to be a great challenge.

4.4 Conclusions

The examples given above show that a company must adapt its policy
with regard to corporate social responsibility to the national social
needs and the local customs that exist in a country.

The examples from China and Brazil are good illustrations of how
different the expectations of those countries are with regard to the
social responsibility Western companies should accept. A good food
supply, decent housing and a basic minimum wage have priority in
China, while philanthropy is not considered to be an issue. In Brazil,
on the other hand, the main themes are reducing social inequality and
companies are also expected to contribute—financially or through
their own personal effort—to reducing the social needs. The gap
between legislation on paper and how it is applied in practice is
smaller in Brazil than it is in China and, in Brazil, independent trade
unions and NGOs have much more influence than in China.

In view of the large differences in social context between countries,
it is very important for international companies to take into consideration the social problems and the political context of a country when
developing their policy with regard to corporate social responsibility
for that particular country. This is also shown by the experiences of

ABN AMRO, Pentascope and Koninklijke Houthandel Wijma. All three companies have had to adapt their company strategy to the local circumstances of the country in which they have invested but, despite this, they have all managed to respect their policy with regard to corporate social responsibility.

Companies that wish to do business in and with foreign countries and wish to accept their corporate social responsibility will benefit if they obtain information beforehand on the attitudes and customs with regard to corporate social responsibility in a certain country. The questionnaire in Box 4.1 is a useful guide for obtaining such information.

1. What is corporate social responsibility mostly associated with in the country concerned?

2. Which social problems have the highest priority in the country? Is the business sector expected to contribute to solving these problems?

3. Which social themes deserve the most attention in relation to the country's exports to other (in particular Western) countries?

4. Does the country's government place higher demands on foreign companies than on local companies? To what extent does the local government invest in responsible environmental and social policies and how well is observance of these policies monitored?

5. Which are the most important groups of stakeholders (for example, the government, customers, the company's own personnel or social organisations) that companies must consider when interpreting corporate social responsibility?

6. Which expectations do these different groups have of the role of the international business sector and that of the local business sector with regard to corporate social responsibility?

7. Is there a well-developed infrastructure of all kinds of social organisation (NGOs and trade unions) attempting to exercise their influence and receiving the opportunity to do so politically? Which organisations are the most influential? What are the usual ways for an international company to communicate with these organisations?

8. Which sociocultural aspects must be taken into consideration when interpreting corporate social responsibility in the country concerned?

Box 4.1 Questionnaire for obtaining country-specific information with regard to corporate social responsibility

5
Chain responsibility in an international context

5.1 The internationalisation of product chains

The increasing globalisation of the economy is making supply chains more and more international. The rich, developed countries in particular are contracting out an increasing percentage of their production to companies in low-wage countries and then selling their finished products in the Western market. They do this to try to stand up to increasing competition based on price, quality and delivery time. There is a risk, though, that, because of this practice, Western companies will become too focused on economic gain and neglect social responsibility and an environmental policy (see, for example, Neef 2004; New and Westbrook 2004; Mamic 2004).

A surprisingly large number of companies are part of international product chains. Whether these concern food products, clothing or electronic appliances, these products are almost all produced in co-

CHAIN RESPONSIBILITY

operation with manufacturers from all over the world. This globalisation of product chains has regularly led to heavy social criticism. Companies have been accused of using child labour or of working in countries that violate fundamental human rights. The criticism was initially especially aimed at large multinational companies, such as Nike and Ikea, but, nowadays, smaller companies that operate internationally also feel pressure to accept chain responsibility.

A growing number of companies of differing sizes, therefore, feel compelled to accept chain responsibility. Chain responsibility is understood to involve urging other companies in the chain to observe (inter)national guidelines and standards with regard to, for example, the environment, human rights, working conditions and integrity. An important, underlying motive for accepting chain responsibility is to avoid liability for problems caused by someone else in the chain. A company's own sense of responsibility also plays a role.

Companies wishing to accept international chain responsibility have to ask themselves many different questions:

- As a company, how far along the chain does your responsibility stretch?

- How can you ensure that the chain partners meet international guidelines and standards?

- How do you make sure that companies observe international guidelines and standards?

In short, how can you, as a company, organise this chain responsibility? To cope with this question, a step-by-step plan which provides practical means to implement corporate social responsibility in international product chains has been developed. This step-by-step plan is explained below and illustrated with company-specific case material.

5.2 Step-by-step plan for global chain responsibility

How a company can organise global chain responsibility depends on a number of variables. First, corporate social responsibility depends on the complexity of the chain in which the company operates. Next, the extent of the ambition to reach certain performance standards is important. In addition, the diversity of the chain and the power of the company in that chain are both influential. These four variables were the key aspects in developing the step-by-step plan shown in Box 5.1.

Step 1. Determine those parts of the product chain for which you can/want to take social responsibility, taking into account the complexity of the chain.

Step 2. Do you focus the product chain on:

   (a) a niche market?

   (b) a mainstream market?

If a, then you can immediately adopt high environmental and social standards.

If b, then it would be better to introduce environmental and social standards which will become stricter over time.

Step 3. Determine the degree of diversity in the product assortment.

   (a) Do you purchase a great diversity of products?

   (b) Are you part of one or a limited number of product chains?

If a, go to Step 4.

If b, go to Step 5.

Step 4. Develop a strategy to ensure that your most important suppliers comply with your ethical code of conduct.

Step 5. Are you able (perhaps in co-operation with a limited number of influential actors) to impose standards on your suppliers and/or other companies in the product chain?

   (a) Yes

   (b) No

If a, then you can take the initiative to organise chain responsibility.

If b, then, preferably, try to seek alliance with an influential organisation or competitors in the same sector who can help organise chain responsibility. If this is not possible, select a limited number of important suppliers with whom co-operation can be established.

Box 5.1 Step-by-step plan for global chain responsibility

5.3 Further elaboration of the step-by-step plan

The step-by-step plan described above will be further explained and illustrated with practical case examples below.

Step 1. Determine those parts of the product chain for which you can/want to take social responsibility, taking into account the complexity of the chain.

An international product chain is usually so complex that it is impossible to get all parts of the chain heading in the same direction. Complex product chains can roughly be divided into two types: chains oriented towards agricultural production (agro-chains) and chains oriented towards industrial production (industrial chains). In general, the agro-chains are more readily surveyed than the industrial chains: the product can be traced back more easily to the primary raw material (Cramer and Klein 2005).

Agro-chains often have the structure of an *egg-timer*. At the beginning and the end of the chain, there are a large number of actors, while in the middle, the number is limited. Recognisable actors at the centre are, for instance, the flower auctions, coffee bean importers and processors, and importers of vegetables and fruit. Those companies often have the power to play a directing role both upstream and downstream in the product chain.

In industrial chains, the structure is often more in the form of a *funnel*. The number of actors decreases along the chain but one cannot speak of concentrated power in the hands of a limited number of actors. For companies close to the market, it is difficult to play a directing role in the whole product chain. The confectionery industry is one of the few chains that still tries to do this actively. This is because this product chain is constantly under societal pressure to improve, for instance, the working conditions of factories in developing countries. The same is true for the chains of toys and baby products. Since industrial chains are usually harder to survey, chain responsibility is often limited to the next link in the chain or, at best, the one after that. Establishing co-operation between the main industrial actors is less easy than in the agro-sector.

To determine the boundaries of chain responsibility, use can be made of the GRI Boundary Protocol, drafted by the Global Reporting Initiative (GRI 2005).[1] GRI recommends that companies take chain responsibility when they have control and/or significant influence in the product chain. Control means having the power to govern the financial and operating policies of an enterprise in order to obtain benefits from its activities. Significant influence is the power to participate in the financial and operating policy decisions of the entity, but without control over these policies. If a company has significant control over activities in the chain and the company also has significant influence, then its responsibility is substantial. A good example of this is the use of express delivery services by ABN AMRO (through its subsidiary Banco Real) in Brazil (see Box 5.2).

In Brazil, Banco Real frequently made use of motorcycle express services. These couriers were not insured and did not wear helmets. The managing director of Banco Real is strongly committed to corporate social responsibility and decided that the bank would only work with express services that took good care of their employees by insuring them and making it compulsory for them to wear helmets. Eventually, ABN AMRO set up an express service together with a local courier which has acquired the exclusive transport rights for Banco Real.

Box 5.2 Subsidiary Banco Real/ABN AMRO

1 The Global Reporting Initiative (GRI) is a multi-stakeholder organisation that has developed international guidelines for reporting the performance of companies with regard to corporate social responsibility.

When a company has influence in the product chain but cannot exert control over it, its responsibility is more limited. Moreover, an important criterion is whether the activities in the chain are risky and can have large negative impacts. The likelier this is, the greater the company's responsibility. However, it remains a decision for the company itself where it draws the boundaries of this responsibility. The credibility of such choices *vis-à-vis* society increases when companies communicate in a transparent way about these choices.

Step 2. Do you focus in the product chain on: (a) a niche market or (b) a mainstream market?

If a, you can immediately adopt high environmental and social standards.

If b, it is better to introduce environmental and social standards that will become stricter over time.

When you strive for very ambitious environmental and social performance standards, it is harder to get suppliers involved. Success is not guaranteed without proper support and training. This implies that, in such circumstances, it is better to start with a limited, carefully selected, group of suppliers. Good examples are Simon Lévelt, a coffee supplier, and Agrofair, a producer of certified fruits (see Boxes 5.3 and 5.4). Through high standards and intensive co-operation with a selected group of producers, these companies distinguish themselves in a niche market. Agrofair also aims to reach the mainstream market with a new product ('citrus').

To reach a mainstream market, initial standards should not be too ambitious. A reasonable number of suppliers should be able to comply with the requirements, which can then gradually become stricter. The MPS Flowers and Plants initiative is a good example of how such an adoption process with a large group of growers can take place (see Box 5.5).

Simon Lévelt is active in the coffee and tea market. The company was established in 1817 and employs 40 people. Around 1976, the former director introduced a policy which could be called corporate social responsibility, although that term did not exist back then. From that time onwards, corporate social responsibility was rooted into the company's culture and the company's core values. Together with the growers of their coffee and tea in developing countries, the company worked to achieve the most sustainable production process possible. This led to the introduction in the Netherlands of biological tea in 1980 and biological coffee in 1984. Simon Lévelt developed guidelines for biological crops, working conditions and honest trade which have been used as a model by other companies. The company's basic principle is that at least 50% of each product category must be biological.

Chain control and chain quality are essential to Simon Lévelt's concept. With its coffee-roasting house and its chain of shops, the company is the link between producers and consumers. The company has direct contact with the producers without any middlemen, which leads to a shorter production chain. The producers receive intensive guidance to improve their production methods. They also receive financial support and even social services. The company uses a number of (high) standards, such as the Max Havelaar and Demeter standards. In order to ensure producers observe these, they are audited by independent consultancies. The costs for such an audit usually have to be met by the producers themselves.

Simon Lévelt feels very strongly about transparency and the traceability of the products. The company recently set up an information system giving consumers full insight into the origin of the coffee and tea. This information can be accessed via Simon Lévelt's website. By doing this, Simon Lévelt hopes to stimulate the consumers to choose sustainable consumption.

Box 5.3 Simon Lévelt

Agrofair was established in 1996 as a company that positioned itself in the niche market of high-quality, certified fruit, such as bananas, pineapples and mangoes. It employs 40 people. The certified products with Fairtrade and biological labels are placed on the market in close co-operation with the producers via a highly integrated chain of suppliers. Fifty per cent of the shares in the Agrofair Europe holding are in the hands of the international producers' co-operative and 50% are in the hands of NGOs such as Solidaridad. Dave Boselie, director of Agrofair Assistance and Development Foundation, explains the co-operation as follows:

> Through this structure, there is a direct trade relationship with the producers and middlemen are unnecessary. We offer the producers an honest and guaranteed price, so that they can build a more secure existence. Producers must observe strict demands regarding the environment and food safety. Monitoring and traceability of the product quality are of the utmost importance. Apart from a cost-covering minimum price, all Fairtrade-certified farmers also receive a premium, which the farmers' organisation invests in local, social development, such as schooling, healthcare and infrastructure.
>
> The Agrofair Assistance and Development Foundation offers producers organisational, financial and technical support. It also offers assistance to meet many certification procedures and to promote exports. In this way, we make it easier for traditional producers to switch to Fairtrade and biological products.

In 2005, Agrofair introduced a new product line, the citrus fruit. Before this product could be placed on the market, Agrofair had to make some strategic choices. Boselie said,

> In order to be able to supply the market throughout the year, the citrus fruit had to be supplied from different parts of the world. The countries chosen are Argentina, South Africa and Morocco. It took some time before we had found interested growers, particularly in Morocco. The next question was how were we going to position the orange. The citrus fruit was less in keeping with the Max Havelaar certification mark than products such as bananas. Furthermore, our competition on the mainstream market had already considerably raised their standards. In order to stay ahead, we developed our own standard which surpasses Fairtrade and Eurepgap. As an extra step, we

Box 5.4 Agrofair (continued over)

chose to offer structural social support in the production countries: in South Africa to black economic empowerment, in Argentina to European market access for small farmers and in Morocco to strengthening the socioeconomic position of farmers and women. At the same time, we chose to focus on the mainstream market with the citrus fruit. We hope to reach a much larger consumer group by marketing the citrus fruit as healthy, tasty and not much more expensive. Within a few years, we will know whether we have been able to achieve this economic expansion.

Box 5.4 (from previous page)

The MPS Foundation (previously known as Environmental Programme Flowers and Plants) was established in 1995. The main drive was the pressure from environmental organisations, particularly in Germany and the Netherlands, to improve the environmental quality of cut flowers. Given that 65% of the worldwide trade in flowers involves the Netherlands, the two main Dutch flower auctions were eager to demonstrate a conscientious response to this criticism. The flowers and plants sector mainly consists of small enterprises which sell through the wholesale trade to the auctions, which then sell the products to wholesalers, supermarkets and small retailers. The approach developed by MPS focuses on gradually improving the sustainability of the production process in close co-operation with a large group of growers. Originally, the approach was limited to environmental requirements, such as the use of energy, water, manure and pesticides. Rapidly, the MPS system was extended to include social aspects: MPS–Socially Qualified. This happened through dialogue with two NGOs and trade unions, with important ILO conventions being used as a starting point.

Box 5.5 MPS Foundation (continued opposite)

MPS is a business-to-business system that is controlled by accreditation bodies. The aim is to co-ordinate the auditing activities, which saves the producers time and money. Within the MPS system, three different levels of environmental and social performance are distinguished related to specific standards. In this way, producers can set their own ambition level. In order to demonstrate compliance with the standards, the producers must have a registration system and be checked by independent auditors. In the Netherlands, 4,500 of the 7,000 growers participate in MPS and 60–65% of the turnover from the auctions comes from certified products. Initially, the MPS was sponsored by the two largest auctions in the Netherlands but, now that many suppliers are involved, the system can cover its own costs.

In 2003, an agreement was reached between the international cut flower sector and an international coalition of social organisations concerning the creation of a new international consumer label: Fair Flowers and Plants (FFP). The social and environmental criteria used are comparable to those of the MPS system. The social organisations will be involved via the review committee, which ensures that the certification organisations involved act correctly. It is the intention that local social organisations will also play a role in the auditing of companies. As a result of the open structure of FFP, various certification organisations are able to market their products under this umbrella. This creates a much wider range of products than if only one certification organisation is used in a country. Furthermore, there will be no competition between the various labels on the consumer market, because FFP is the only label.

Box 5.5 (from previous page)

Step 3. Determine the degree of diversity in the product assortment.
(a) Do you purchase a great diversity of products? (b) Are you part of
one or a limited number of product chains?

If a, go to Step 4.

If b, go to Step 5.

Step 4. Develop a strategy to ensure that your most important
suppliers comply with your ethical code of conduct.

A company that purchases a great variety of products has to deal with
many very different and sometimes changing suppliers. Examples are
ABN AMRO (a bank), De Bijenkorf (a large retailer) and Merison (a
supplier of non-food products to supermarkets). For such companies
it is an almost impossible endeavour to have a clear overview of all
their product chains, let alone to audit the social responsibility of
those in the chains. How can they then still take chain responsibility
in such a situation? All the three companies mentioned above began
the process by formulating an ethical code of conduct and, based on
that, developed a checklist for a sustainable purchase policy. Next they
discussed the code of conduct with their main suppliers and asked
them to sign the code and complete the checklist. Checking the relia-
bility of the answers turned out to be a time-consuming (and thus
costly) business. In order to still incorporate some monitoring, the
three companies mentioned above began to select a number of suppliers, particularly those who carry out potentially risky or socially unde-
sirable activities and those who are clearly related to the brand image
of the company. By focusing on a selected group of suppliers, companies are able to check through third parties (such as certification bodies) whether supply chain partners comply with the desired code of
conduct. A few examples are given in Boxes 5.6–5.8 to illustrate this
approach.

De Bijenkorf is a well-known chain of department stores in the Netherlands and is a part of Vendex/BB. The company has 13 branches and a total of 3,700 employees. Pieter Huygens, merchandise controller, is responsible for the qualitative and quantitative aspects of De Bijenkorf's flow of goods and he explains the policy with regard to corporate social responsibility as follows: 'As a result of the variety in products, it is very difficult for us to visualise the chain responsibility'. To show us how difficult that is, he produced a letter from an NGO which asked for the origin of a pair of leather gloves which were purchased at De Bijenkorf. The phrasing of the question was so detailed that they even wanted to know the tanning process and the living conditions of the goat from which the leather came. 'We felt that the letter had to be answered seriously, since NGOs are, after all, the thermometers of society.'

Huygens continued,

> We are working very hard to set up a sustainable purchase policy, but that is not easy. De Bijenkorf has 300,000 articles in its range, which come from more than 1,000 different suppliers throughout the entire world. These products are purchased by 30 purchasers and a large number of agents. Since the tasks of these purchasers vary, they cannot build up specific know-how of a certain product group. Although De Bijenkorf is centrally controlled, matters that do not have a support base—in this case, from the purchaser—are difficult to get off of the ground. This is typical of De Bijenkorf's culture.

How has De Bijenkorf developed its sustainable purchase policy? Huygens:

> In 2003, we started drawing up a company code which states how De Bijenkorf wishes to interpret corporate social responsibility. Next, we had to set priorities. In order to define the issue and approach things step by step, we started with the products that we sell under our own label. De Bijenkorf can be called to account about these products first, because it is De Bijenkorf itself that is responsible for the labels. We have produced a manual with a checklist for these products for the purchasing department, which is based on the company code. The checklist contains tips, checks, questions and actions concerning corporate social responsibility for every phase of the purchasing process. The manual is to be used as a reference. The intention is that

Box 5.6 De Bijenkorf (continued over)

every De Bijenkorf purchaser will use this information as a basis for his or her actions. The first positive experiences have now been gained with this manual. It is important to make good agreements concerning who is responsible for what during the process and to keep the documentation up to date. A support base is necessary for the process to be successful and we create this by holding various sessions within the organisation. This works very well.

Our biggest problem was checking the completed checklists. How could we ensure that our suppliers answer them honestly? It was impossible for us to monitor and audit our suppliers ourselves.'

Via contacts with the Association of Chain Stores in Textiles (Vereniging van Grootwinkelbedrijven in Textiel [VGT]), De Bijenkorf has become affiliated with the Business Social Compliance Initiative (BSCI). 'This organisation is in line with our needs', said Huygens. BSCI is an initiative linked to the retail sector (see www.bsci-eu.org). Through this initiative, audits are co-ordinated at suppliers on the basis of certain quality standards. These audits are carried out by independent certification bodies.

Box 5.6 (from previous page)

ABN AMRO has approximately 98,000 employees and its home markets are the Netherlands, Brazil and the United States. Dennis de Jong, senior vice president of procurement development, explains the recently developed sustainable procurement policy as follows.

> Our Global Procurement Sustainability Policy is the basis of the policy. Linked to this is a code of conduct for suppliers and 20 standard selection criteria with regard to corporate social responsibility. We are currently in the process of having the (large) suppliers sign the code of conduct. The 20 selection criteria are already being included as standard in every tender together with other criteria, such as quality, delivery reliability and price. We can also make additional demands for specific products. There is no contractual obligation to observe the code of conduct, but if it is not observed we will certainly engage in dialogue with the supplier concerned. We are considering auditing a selection of suppliers in the future, although it is not yet clear how we can best do this but we are using the experiences gained by our bank in Brazil. The integration of corporate social responsibility in the procurement process has been reasonably smooth and commitment from the top management with regard to corporate social responsibility has clearly helped with this.

Mark Hanhart, who was also involved in the development of the sustainable procurement policy, adds:

> Sometimes, suppliers find it difficult to respond each time to slightly different questions from their customers. We, therefore, first compared a large number of standards that other companies already use and then expressed our policy as much as possible using commonly used standards. It would be better, of course, if we worked towards one joint, uniform standard for similar customers, but we have not reached that stage yet.
>
> The progress of the sustainable procurement policy is determined based on the number of suppliers who have signed the code. The number of tenders that include the sustainability criteria will also be counted. Certain themes receive a higher priority in a certain region. For example, in the United States diversity is a central theme, and in Brazil the employment conditions for the personnel of the suppliers are important. The reactions from suppliers in our home markets have been predominantly positive and some have even seen it as a wake-up call.

Box 5.7 ABN AMRO

Merison is a wholesaler in non-food articles for retailers in the Netherlands (including Albert Heijn). The company was established in 1896 and employs 35 people. Monique Berings, co-ordinator for the policy with regard to corporate social responsibility, explains the development of this policy within Merison:

There was already a code of conduct when I began working at Merison in 2001. This code particularly emphasised the working conditions and was drawn up in 1997 together with social organisations. The company also participated in a number of social projects. My task was to implement the code of conduct in the purchase policy. It was decided to ask approximately 20 suppliers each year to sign the code of conduct and to have an audit carried out by an independent certification institute (SGS) for four or five products. Where permitted, a Merison employee would be present during the audit. This turned out to be very informative. In the job description of Merison's purchasers, it was included that they had to get a number of suppliers to sign the code of conduct each year. At the start of the year, it was agreed which suppliers that would be—a decision based on the contribution to the turnover of Merison and the amount of risk involved with the product.

Now, by the middle of 2005, more than three-quarters of the most important suppliers have signed the code and 40% have been audited. The suppliers do not (currently) have to pay for the audit. Instead, Merison makes €10,000–15,000 available each year for the audits. The purchasing manager is responsible for achieving the targets. I, myself, assist the Purchasing Department in carrying out this process.

The basic product range, which is supplied by some 100 companies, has now mostly been mapped out. Most of these products are reasonably well or quite well in line with the code of conduct. A correction plan has been drawn up for products where this is not the case. Building relationships with suppliers of promotional products and seasonal products is quite a bit more difficult. These relationships are usually shorter and often change. Since we are often only a small customer to them, they are not keen to sign the code or have an audit carried out. Nevertheless, Merison tries to come to an agreement with them, particularly if it is a risky product.

Suppliers are usually pleased with the audit afterwards. After the audit has been carried out, the suppliers sometimes ask if they can receive a certificate, but such a certificate has little meaning if it only concerns Merison's deliveries. The willingness to co-operate and even pay increases when the results are certified by a generic audit system.

Box 5.8 Merison

Step 5: Are you able (perhaps in co-operation with a limited number of influential actors) to impose standards on your suppliers and/or other companies in the product chain? (a) Yes (b) No

If a, you can take the initiative to organise chain responsibility.

If b, look for co-operation, preferably with an influential organisation or competitors in the same sector who can help organise chain responsibility. If this is not possible, select a limited number of important suppliers with whom co-operation can be established.

The extent to which the company taking the initiative has the power to place preconditions on its suppliers is crucial to the organisation of chain responsibility. If the company has that kind of power, it can draw up a code of conduct and purchase priorities based on the code and then impose these on suppliers. In order to ensure that the suppliers are able to meet these demands, it is important to offer support when implementing and also when monitoring (auditing) the requirements. In the example of MPS, the two auctions had so much power that they were able to involve the growers in the process. To organise the process, a separate organisation (MPS) was established at arm's length. Also Simon Lévelt and Agrofair had sufficient control and influence over their suppliers to ensure their commitment.

In many cases, companies do not have such a large influence and control that they can direct the product chain. Good examples are Difrax (producer of baby products; see Box 5.9) and AXA Stenman (whose products include locks for windows, doors and bicycles; see Box 5.10).

Both work a lot with suppliers in China. Part of the work is contracted out to more producers there, which makes monitoring even more difficult. Since Difrax and AXA Stenman are not the suppliers' only customers, the suppliers can, in principle, refuse the purchase conditions if they are too difficult for them to comply with.

Nevertheless, in order to commit the suppliers to certain social conditions, Difrax has visited all 40 suppliers and presented its purchase conditions to them (see Box 5.9). Fortunately, all of them were prepared to co-operate. At the same time, Difrax affiliated with the Business Social Compliance Initiative, just as De Bijenkorf had done. The

Difrax develops and sells baby products for the Dutch and European markets. It has 25 employees. The products are developed in the Netherlands and produced in various countries including China, Thailand and Spain. Viviënne van Eijkelenborg, Managing Director of Difrax, explains the company's policy as follows:

> A few years ago, I took the initiative to develop a policy on corporate social responsibility for my own company and our suppliers. This policy had to be monitored and had to lead to a quality guarantee for our products. This turned out to be a hell of a job. Our biggest problem was to introduce one uniform policy among our suppliers. Of course, we were willing to take into account the diversity of suppliers and their different locations, but the question was how. As a first step, we asked our suppliers to fill in a questionnaire, which contained questions about the activities already carried out by suppliers in terms of corporate social responsibility. The answer frequently heard was that they had already been audited many times over and were graded as 'sufficient'. This answer did not help us much. Therefore, we asked our five largest suppliers in Thailand, China and Taiwan to fill in a more specific questionnaire. We used the questionnaire of a certification body as a basis, and then adapted, shortened and applied it. Central questions were: What are the contents of the legislation in your country? With which standards do you need to comply and are you doing that? The information acquired provided some insight into the situation, although some companies filled in the questionnaire more professionally than others. Moreover, we make use of some 30–40 smaller suppliers. Some of them have a Western management which complies with high international standards. Others, however, are less keen to fully adopt international codes of conduct.
>
> On the basis of the first round of conversations, we concluded that suppliers are already doing a lot of things. Some factories are audited every week or month by various other organisations, while others are not audited at all. The factories are familiar with all kinds of international code of conduct in the field of corporate social responsibility and a number of them comply with some of the guidelines. However, one company was relying on one particular code, and a second on another code. It was a real mess. The question was how Difrax could proceed. We concluded that we are too small to carry out this task alone. Organising audits and monitoring ourselves would be too complex and time-consuming. We sought simple solutions.

Box 5.9 Difrax (continued opposite)

> Therefore, we made an inventory of organisations which could support us. The International Council for Toy Industries (ICTI)—the sector organisation for the toy industries—was an option. The advantage of this organisation was that it was in direct contact with the appropriate suppliers. However, ICTI did not cover all of our products. Another option was to co-operate with SGS, a certification body. This is a reliable organisation, but we were searching for a broader initiative. Finally, we got in touch with BSCI (Business Social Compliance Initiative), an initiative linked to the retail sector. Through this initiative, audits are co-ordinated at suppliers on the basis of certain quality standards. These audits are carried out by independent certification bodies.
>
> The costs of the audits are often paid by the suppliers. The advantage for Difrax of joining BSCI is that we do not need to reinvent the wheel and can make use of the auditing knowledge of BSCI. Moreover, by combining forces, repeat visits to suppliers by different auditors can be avoided. Recently, we have informed our suppliers that we are participating in the BSCI. The objective is to start the audits in 2005 and to have all 40 factories approved by the end of 2006.

Box 5.9 (from previous page)

advantage for Difrax is that, as a small company, it does not have to reinvent the wheel itself and it can make use of the audit knowledge of the BSCI. The combining of forces also prevents suppliers from being interviewed by different auditors each time one of their customers requests an audit.

Unfortunately, there are only a few sector-specific or broader initiatives that can provide support in organising chain responsibility in a similar way as in the case of Difrax. For instance, AXA Stenman encountered difficulties in finding a suitable initiative. The appropriate Dutch sector organisation was not active with corporate social responsibility in international product chains. BSCI was not an alternative, because AXA Stenman did not belong to the primary target group. Finally, the company has chosen to use SA8000 (an international labour standard) as a basis for the AXA Social Charter. The com-

AXA Stenman was established in 1901. The company develops, manufactures and sells locks, handles and hinges for doors and windows, and bicycle accessories and employs approximately 250 people. AXA Stenman is making increasing use of products from the Far East for the production of its own components. It is also becoming more common for finished products that have been developed by AXA Stenman to be manufactured in the Far East. This was a reason for the management to start paying attention to chain responsibility in an international context in 2003. Caspar Vedral, the Health, Safety and Environment Co-ordinator, explains how the process went:

> First, we attempted to form an image of the international standards and guidelines with regard to corporate social responsibility. Next, the OECD *Guidelines for Multinational Enterprises* were used as a start and these were transformed into a questionnaire which was presented to the management team. All the members of the team were requested to indicate which subjects should receive the highest priority within the framework of a sustainable purchase policy. The subjects that scored high were respecting human rights, refraining from discrimination, the right to free trade unions, fighting corruption and protecting the interests of consumers. The criteria for the purchase policy could be derived from these results. The question then was how we could implement this purchase policy. Should we do it all ourselves or join an existing initiative? The answer appeared to be obvious. We were not going to reinvent the wheel and, therefore, we wanted to join an initiative that suited us. Firstly, we sounded out our branch organisation FME on the subject, but they were not concerned with corporate social responsibility in an international context at that time. Another possibility was to join the Business Social Compliance Initiative (BSCI). Two other participants of the programme were already affiliated to this initiative. Unfortunately, our company was not part of BSCI's primary target group.
>
> The need to continue appeared to be much closer to home than we thought. When our sister company in France entered into a long-term agreement with a department store chain, the purchase conditions contained specific demands that related to corporate social responsibility. This company's contract contained an article that concerned a social charter, which also determined the supplier's chain responsibility. This social charter was mainly based on SA8000.

Box 5.10 AXA Stenman (continued opposite)

Soon afterwards, an interview with a Dutch colleague from the bicycle sector appeared in an international bicycle magazine. In this article, he called on colleagues within the sector to set up a Purchase Charter so that the bicycle industry as a whole could give shape to its joint social responsibility. This call offered us the possibility to make a start with partners from the sector. The following was decided:

1. We have chosen SA8000 as the basis for the AXA Social Charter.

2. We are going to discuss the AXA Social Charter with our agent in the Far East. Together, we will look to see how the charter can be introduced step by step at the existing suppliers

3. New, potential suppliers in the Far East will receive the AXA Social Charter immediately, accompanied by a simple questionnaire which the supplier must complete

4. In order to find out the supplier's status, we expect to receive a plan of improvement for important deviations from the charter. Support and monitoring will be the local agent's responsibility

5. We wish to give the above a broader base through contacts with our colleagues in the sector

Box 5.10 (from previous page)

pany is now looking to co-operate with interested parties in the sector to broaden the initiative.

The Cookie Company, which markets licensed baby and children's products, in itself does not have enough power in the chain to force certain social values on suppliers, but this company could take advantage of the attitude of one of its most important licensers, Walt Disney (see Box 5.11). By doing so, it was possible to have the Walt Disney standard implemented at its suppliers in a short space of time, but, as Box 5.11 shows, it was not all without problems.

The Cookie Company develops products in the Netherlands and has them manufactured mostly in China. The company has 25 employees. This company also has to contend with the problem of how a small company can organise chain responsibility. Instead of affiliating with an organisation such as BSCI, it gave shape to its chain responsibility under the flag of its most important licenser, Walt Disney. Sabine Hulsman, director of The Cookie Company, explains this as follows:

> For the last few years, Disney did not permit production of its brands before the production company had become certified according to its standards. As a result, the process of corporate social responsibility developed very rapidly. The Cookie Company has found more than half of its 25 suppliers (mostly situated in China) prepared to co-operate with the audit demanded by Disney. The fact that Disney is an important customer to them contributed to this. After agreeing to co-operate, we have helped these suppliers to prepare for the audit. Disney will only pay for the audit (approximately €1,500) if the company successfully passes the audit the first time. If the company does not meet the standard the first time, then all the costs—also for the following audits—have to be met by The Cookie Company. This was the case for all suppliers. The Cookie Company also paid for some of the necessary alterations within the production companies.
>
> Many companies did not even receive the OK status after the second audit. This was often because they did not understand the instructions. In order to solve this problem, a Chinese employee from our company visited these companies and went through all the problem areas with them. These concerned basic rights, such as minimum pay, paid overtime, having an emergency plan and displaying this in a visible location in the factory, and the rights and obligations with regard to safety. This last point meant, for example, indicating emergency exits and mentioning the safety certificates the company has. The problem was usually that these problem areas were not properly registered or that certain documents were missing. These omissions were often easy to put right: for example, simple things, such as obtaining a certificate from the fire brigade and registering things better in the administration. Five suppliers approached by The Cookie Company were not prepared to be open with regard to the number of employees and their working hours and The Cookie Company now no longer does any business with these companies. My experience is that we often think a bit too arrogantly about corporate social responsibility. In China, the basic rights are not even properly organised. Making those rights visible would, in itself, be an enormous advantage.

Box 5.11 The Cookie Company

5.4 Conclusions

The ways in which companies can organise chain responsibility in an international context turn out to vary substantially. It depends on the complexity and diversity of the product chain, the level of ambition set and the power of the company in the chain. The step-by-step plan described above offers guidance in choosing an appropriate way of organising chain responsibility. With the help of this step-by-step plan, companies can put their corporate social responsibility into practice in a proactive way.

In implementing global chain responsibility, companies will encounter a variety of problems.

First, establishing co-operation with suppliers may be difficult. This can be for various reasons. Some suppliers may simply reject the requirements of a customer if the customer is not seen as being important enough. The only thing the customer can do in such cases is to break off the relationship. Moreover, as frequently happens, suppliers do not really understand the environmental or social requirements set by their customers. For instance, overtime in Asian countries is looked on completely differently in comparison to many Western countries. Employees in many Asian countries want to work as much overtime as possible in order to earn extra money. International standards, however, often restrict overtime to 12 hours per week. On the other hand, suppliers are often forced by their customers to deliver products quickly and at a low price. That policy is contrary to the social demands set within the framework of corporate social responsibility.

Second, checking whether suppliers comply with requirements set is a complicated task. For companies with a diverse product range, monitoring all their purchased products is impossible. Therefore, such companies usually limit themselves to products that may lead to risks or are strategically important. When companies operate in only a very limited number of product chains, it is easier to monitor compliance, although it is still not foolproof. Often, recognised certification bodies are hired to audit companies. When they are well trained and knowledgeable about the local situation, the monitoring can be adequate. In this respect, experiences with the capabilities of auditors differ. Moreover, the quality of the work by auditors depends on the degree to which they are able to include the views of employees and local stakeholders. This is a weak point in many audits. The need to instruct suppliers properly about the improvements to be made is also often overlooked. As a result, the same omissions are found in subsequent audits.

Who is responsible for the costs of the audits and the corrective actions is also a delicate issue. In principle, these costs should be carried by the suppliers who will, in turn, cover them in the price of their products. The willingness to pay for a certificate increases if the certificate is valid more widely. In this way, a supplier can avoid being visited each week by a different auditor. A benefit to suppliers is that customers who adopt high environmental and social standards are generally more loyal. They organise training to ensure that their suppliers can meet the requirements set and also support them in carrying out corrective actions. They often invest in local social development: for instance, in schooling, healthcare and infrastructure.

Third, transparency is needed in order to gain acceptance of the company's approach by both internal and external stakeholders. To enhance transparency concerning the compliance of suppliers with international guidelines and standards, it is recommended that a company reports about: (a) the quality and working methods of the auditing body (including the involvement of the local community and employees); (b) the results of the audit and the corrective actions proposed (including a plan for when these actions should be completed); (c) the issues where the existing policy deviates from international guidelines and standards and why; and (d) the manner in which the organisation supports their suppliers financially or otherwise in conforming to the requirements set.

Fourth, the playing field is changing now that companies in the mainstream market are giving increasing attention to corporate social responsibility. Companies that once distinguished themselves in niche markets with socially responsible Fairtrade products are now beginning to lose their monopoly position. Therefore, the question of whether they have to raise the requirements for corporate social responsibility even higher is becoming increasingly important to them. The tendency is to do so and, at the same time, to focus more on the mainstream market.

Finally, the reality is that many companies cannot organise global chain responsibility by themselves. They simply lack sufficient power in the chain. Many also see reinventing the wheel as a waste of time and energy. Combining forces is therefore urgently needed. This could be co-operation within a cross-sector initiative (such as the Business Social Compliance Initiative), a sector-specific initiative (such as in the clothing industry or the agricultural sector), a multi-stakeholder organisation (such as Fair Flowers and Plants and Agro-fair) or co-operation with an important customer (such as Walt Disney). Unfortunately, the number of such initiatives remains limited. In order for global chain responsibility to really take off, joining forces should be the top priority.

6

The contribution made by international companies to the local economy of developing countries

6.1 The societal effects of foreign investment

The previous chapters dealt with companies giving shape, in a socially responsible way, to policy with regard to people, planet and profit within their own companies and in the chain. The contribution that international companies can make to economic development and combating poverty in the countries in which they invest has hardly been looked at. However, this subject deserves attention from companies who want to accept corporate social responsibility in an international context. Particularly relevant is the question of to what extent

foreign investors have a positive or negative influence on the local economy, in particular in developing countries.

In the 1970s, public opinion of this was sceptical. Social organisations and governments often considered private investors to be imperialists who were only looking for profit and not for an improvement in the economic structure of the host country. Nowadays, there is a more balanced opinion of the role of international business in developing countries. Plus points are, for example, the financial contribution to the national treasury (via taxation), the positive effect on direct and indirect employment and the transfer of science and technology. On the other hand, there are also critical points. For example, international companies can drive away local entrepreneurs, abuse their powerful position in the market and maintain low social and environmental standards. How people weigh up these positive and negative points depends a lot on their opinion of globalisation and the role of international business in this. Supporters of globalisation emphasise the positive points and opponents emphasise the negative points.

The most important question for international companies themselves is: what concrete contribution can they be expected to make to the local economy both in general and in specific countries? It is often not clear in advance whether and, if so, in which respect, a foreign investment is beneficial to the economic development of the host country. Which criteria can a company use when considering this question? This is the topic of discussion below.

6.2 Views based on theory

Relying purely on literature, it is difficult to find an answer to the question of what contribution can be expected from international business to strengthen the local economy, particularly in developing countries. A great deal of literature focuses on the pros and cons of globalisation and the role of the international business sector in this. The different views on this theme appear to vary greatly. Critics are extremely negative. Viviane Forrester, for example, suggests in her book *De Terreur van de Globalisering* (*The Terror of Globalisation*) that liberalism leads to a growing concentration of power with multinationals. They only care about their shareholders and play ruthless poker with their employees throughout the world. Governments have little say in what these mega-companies do (Forrester 2001). Along the same lines, Naomi Klein (2000) expresses her concern over the 'McDonaldisation' of the world.

Economist and journalist Noreena Hertz is also very critical, particularly of the undemocratic way in which the international business sector operates globally. She wonders what the net result of global capitalisation is, in a world where the economic well-being and the physical safety of people is mainly determined by the strategies and actions of international financial investors and multinational companies. Hertz agrees, though, that the business sector reacts to external pressure and is increasingly taking responsibility for solving social problems. She concludes that an increasing number of companies are accepting more responsibility during the earliest phase of the 'Silent

Takeover'. Many of the tasks that governments are becoming less able to effectively fulfil and many of the responsibilities which they are increasingly unable to comply with are being slowly taken over, not only by individual businessmen, but also by the companies themselves (Hertz 2002).

John Cavanagh and Jerry Mander express their feelings using similar words to those of Hertz:

> Where corporate globalists see the spread of democracy and vibrant market economies, citizen movements see the power to govern shifting away from people and communities to financial speculators and global corporations dedicated to the pursuit of short-term profit in disregard of all human and natural concerns. They see corporations replacing democracies of people with democracies of money, replacing self-organising markets with centrally planned corporate economies, and replacing diverse cultures with cultures of greed and materialism (Cavanagh and Mander 2002).

There are also clear supporters of globalisation. A good representative of these is Johan Norberg. In his book *Long Live Globalisation*, he says that both the poor and the rich profit from a free market. Never in history have hunger and poverty disappeared so quickly as in the last few years thanks to the free market. The misery has only increased in countries with a closed economy. Therefore, it is not globalisation that is the problem, but the lack of it. According to Norberg, it is not free trade that has caused hunger and suffering in the developing world, but the lack of it (Norberg 2002). Kofi Annan, Secretary-General of the United Nations, adds: 'The poor are not poor as a result of too much globalisation, but rather of too little globalisation.'[1]

Most authors have a balanced opinion towards the question of globalisation and the role of the international business sector in this (see Naert and Coppieters 2000). For example, according to George Soros, free competition on a global scale has given creative and pro-

---

1   Kofi Annan, speech during the Millennium Forum, 22 May 2000.

fessional talent a chance to develop and has accelerated technical innovations, but globalisation also has negative sides to it. First, many people, particularly in less developed countries, have suffered damage as a result of globalisation without receiving support from the social safety net, and others have been locked out of global markets. Second, globalisation has led to a disruption in the existing balance of the division of resources between private goods and collective goods. Markets are good at creating wealth, but are not intended to be concerned with other social needs. Carelessly striving for profit may damage the environment and clash with other social values. Third, global markets are sensitive to crises, according to Soros (2002).

Scientific studies of the contribution made by foreign investors to the economic productivity of host countries do not lead to unequivocal conclusions either. Studies based on the economic analyses of various sectors generally present a positive image. Studies that use panel data, on the other hand, show negative or non-significant effects (Fortanier 2004). It is clear, though, that some developing countries have been able to enter the world market and compete with the West. Since 1980, their share of exports has grown from 25% to more than 80% (World Bank 2002). On the other hand, other developing countries have become increasingly marginalised in the world economy and have seen their income fall and their poverty increase. Important reasons for this are the trade barriers that the industrialised countries have set up and the measures they take in order to protect the agricultural sector (including import and income subsidies). There is also sometimes a lack of basic conditions in developing countries, such as political stability, a co-ordinating and transparent legal system, a good financial policy, a strong local economy and a good physical and social infrastructure (such as healthcare and education).

The negative effects of globalisation have led to widespread protest from a variety of non-governmental organisations (NGOs): namely, anti-globalists and proponents of the alter-globalisation movement. According to the economist van Liemt, there are two types of anti-globalist, who differ in the way in which they protest and the objectives they strive for. He suggests, however, that:

they share the same world-view. In their eyes, the legiti-
mate interests of many people are pushed aside for the
power of large companies who allow the short-term pur-
suit of profit to prevail over the fulfilment of important
needs, such as caring for the environment, protecting
human rights and combating poverty. They believe that
global governance and the way in which international
institutions function mostly serve the interests of large
companies. They, therefore, consider it their duty to offer
an opposing view to counterbalance the increasingly
more effective lobbying from the business sector (van
Liemt 2002).

According to these people, as long as access for developing countries
to the world market is hindered by the protective measures of Western
countries, it will not be possible to have a level playing field.

The attitude of these NGOs towards the business sector varies.
According to Barrrez (2001), the gap between many anti-globalists
and the business world runs very deep. Even Noreena Hertz claims
that anti-globalists are critical of all the good intentions of multinational companies. According to her, they consider it too large a risk to
assume that what is good for the business world is also good for society. While some possibly welcome the recent attempts by various
companies to tackle a number of the shortcomings of the system and
to contribute to the social climate, they are inclined to see these
actions as a flattering picture of things or as public relations and they
remain sceptical of the companies' motives (Hertz 2002).

However, there is also a large group of NGOs that have a positive
attitude towards the international business sector with regard to corporate social responsibility. Most of them wish to have better information and a more transparent and democratic decision-making
process. They wish to influence the decision-making through dialogue (van Liemt 2002). Through consultation with the business sector, this group of NGOs wishes to take steps towards corporate social
responsibility. Such alliances are also called partnerships.

There is wide support for such partnerships. For example, it was
one of the most important topics of conversation during the World

Summit on Sustainable Development in Johannesburg in 2002. The Green Paper published by the European Commission in 2001 about encouraging corporate social responsibility also argued in favour of close partnerships

6.3 Views based on practice

The insight found in literature has, unfortunately, turned out to be of too general a nature to offer companies a practical guide. Literature contains many macro-considerations. The specific characteristics of a certain host country and of the type of foreign investment are not included and this is exactly the kind of information that is important at a micro level to get a clear idea of the concrete contribution that the international business sector makes at this current time. In order to gain some insight into this, representatives from nine companies have provided this information. These companies have branches in developing countries and could contribute their own experiences. Two representatives from the Dutch government and four from NGOs also provided their view on the contribution made by multinational companies to the local economy of developing countries.[2] What insight did this information give us?

In particular, international companies can stimulate the local economy of developing countries by their contribution to economic growth and economic development. International companies can also contribute to the local economy via sponsoring and charity activities (see Box 6.1).

---

2 This research was carried out by Rosalie Porte within the framework of her business administration study at the Radboud Universiteit in Nijmegen, Netherlands. She co-wrote with J. Cramer 'The Contribution of Multinational Organisations to the Local Economy of Developing Countries', in Cramer *et al.* 2004: 37-51.

The contribution to economic growth is shown in the growth of the gross national product of a country. The growth is promoted by:

1. The employment opportunities that multinational companies create directly or indirectly and the wages they pay their employees

2. The taxation they pay

3. The profits they make that are partly invested back into the local company

The contribution to economic growth gains shape through:

1. Helping to open up under-developed regions by, for example, improving the physical infrastructure or the communication possibilities

2. Transferring scientific and technological knowledge

3. Positively influencing local activities

Sponsoring means carrying out activities that serve the interests of those being sponsored as well as the interests of the company.

Charity means carrying out activities where the interests of the company play no role.

Box 6.1 Ways to stimulate the local economy of developing countries

From interviews with the representatives of the nine international companies, it became apparent that they are generally positive about the contribution their companies make to the local economy of developing countries. In their opinion, they contribute in several ways. They pay wages that, generally, are higher than those paid by local companies. They have a positive effect on employment. They pay taxation as one of their legal responsibilities and they (re)invest their profits into the country concerned if this is necessary or attractive for them to do so. Some companies actively encourage the government to improve the infrastructure and communication systems and, in this way, help to open up under-developed regions. When deemed necessary, they do this themselves. The companies also transfer knowledge. This happens, for example, by working together with local companies to improve the efficiency of their production process and the quality of their products (see the example of Friesland Foods in Box 6.2).

Friesland Foods supports local dairy farmers in countries where it has production locations. The subsidiary Dutch Lady in Vietnam, for example, has helped to set up a dairy farm in Vietnam in co-operation with local dairy farmers and the Vietnamese government. The technical support included breeding strong and productive dairy cows and building an experimental farm with training and advice facilities. Furthermore, Dutch Lady Vietnam has developed a credit system. The cattle farmer buys cows from Dutch Lady Vietnam and pays for them with the milk money. This enables the cattle farmer to continue to invest and to increase his income. The dairy farmers also receive a bonus if they meet all the hygiene requirements. Finally, Dutch Lady Vietnam has arranged for the milk to be transported from the farm to a cooling centre as quickly as possible and from there to the factory. Thanks to this support, many dairy farmers are now able to independently provide farm milk of a good quality which guarantees them an income.

Box 6.2 The transfer of knowledge by Friesland Foods

Representatives from the nine companies also believe that their presence is good for local activities and that they generate local opportunities.

The degree to which companies are active with regard to sponsoring and charity work differs between companies. Sponsoring is preferred over charity, because the interests of the company are also served. Some good examples of sponsoring that serve both the interests of the person or subject being sponsored and the company that does the sponsoring are the activities of the Shell Foundation (see Box 6.3) and of Simon Lévelt (Box 6.4).

The Shell Foundation is a foundation for which Shell provided the capital, but which operates independently from Shell. Since 2000, the Shell Foundation has had a policy of supporting the financial and economic position of poorer communities. Within the framework of this pro-poor enterprise development policy, the Shell Foundation has set up four projects in co-operation with the local population. These projects are: (1) sustainable solutions for interior air pollution in Guatemala and India caused by polluting cooking equipment; (2) promoting the use of solar energy in the houses of some of the poorer communities in India; (3) providing support through the set-up of a social microfinance system in India; and (4) providing support to small and medium-sized business activities aimed at providing energy for poorer communities in Uganda and South Africa (www.shellfoundation.org).

Box 6.3 The sponsoring activities of the Shell Foundation

For more than 15 years, Simon Lévelt has been supporting its coffee suppliers in Mexico, Peru, Nicaragua and Uganda and its tea suppliers in India. For example, in the north of Peru, Simon Lévelt has sponsored the construction of a drying facility for (biological) coffee and the decoration of a testing room at Pronatur-Aproeco, an organisation of 1,300 small coffee growers. There is a similar project in Uganda for small growers of biological coffee in the Busheny district. Thanks to this form of sponsoring, the coffee growers are able to deliver a better-quality product, which strengthens their market position and they receive a higher price for their coffee. Simon Lévelt's advantage is that it is supplied with good-quality coffee.

Box 6.4 The sponsoring activities of Simon Lévelt

The representatives from the social organisations were more sceptical of the contribution made by international companies. According to them, companies could do much more than they are currently doing. In the first place, international companies should ensure that they do not damage the local economy in developing countries and should, therefore, choose a no-harm policy. For this purpose, companies should determine the effects of their activities on the socioeconomic development of the host country. According to the interviewed social organisations, such a socioeconomic effect measurement should be carried out before a company invests in a country and should be repeated once it has established itself. At the moment, such effect measurements are carried out only sporadically.

The social organisations interviewed had their doubts about whether the international companies actually pay a liveable wage and they were of the opinion that more local employees could be employed, also in higher positions. They found the negative influence that international companies have on the local activities in a developing country worrying. They also expressed their distrust of the way in which international companies negotiate with the governments of developing countries over their tax rates. There is also doubt about the extent of the transfer of scientific and technological knowledge. Finally, the social organisations did not consider the opening-up of under-developed regions to be the task of international companies.

The two government representatives were less sceptical of the contribution made by international companies. According to them, these companies should, above all, ensure that they use the OECD guidelines for multinational enterprises as a guide for the way in which they operate. Furthermore, they indicated that some aspects are mainly a responsibility of the local government, such as the development of under-developed regions and certain forms of sponsoring.

Both the social organisations and the government representatives were positive about the various sponsoring and charity activities of international companies in developing countries. They prefer sponsoring activities, because these are linked to the company's core activities.

According to the social organisations, more attention should be given to improving the income position of the poorest communities. This must be inspired by the bottom-of-the-pyramid philosophy developed by Stuart Hart (2005) and elaborated by C.K. Prahalad (2005). According to these management gurus, there are large market opportunities at the bottom of the spending pyramid. By introducing products that are adapted to meet the needs and possibilities of the poorest people, their living conditions can be considerably improved. The reduction in costs and alterations to the processes and distribution methods which are necessary to do this lead to interesting innovations which can also be of value to the top of the spending pyramid. Examples of initiatives that have already proved their added value are the setting-up of a savings system (see the example of the Mexican cement factory, CEMEX, in Box 6.5) and the issuing of microfinance to the poorest people, who can then use the money to set up a small company (see the example of microfinance lending in Box 6.6).

The Mexican cement factory CEMEX set up a savings and credit system for poor Mexicans who were not able to pay for the construction of their houses. A franchise system was set up where the local Mexicans could buy basic products for building their houses. Gradually, they were able to pay for these via the savings system. The local Mexicans formed small groups (cells), which were jointly responsible for paying the money to CEMEX; the social control that was created as a result worked out positively. This approach has helped 75,000 families. This initiative also had a positive effect on CEMEX: the demand for cement has increased.

Box 6.5 The Mexican cement factory CEMEX

The first bank for microfinance was created as a result of an initiative by Muhammad Yunus. In his country of birth, Bangladesh, he wanted to buy a couple of bamboo stools from a woman in a small village. The woman told him that she only made them for the man who gave her the bamboo. He gave her the bamboo on the condition that she would only sell to him for a price fixed by him. As a result, the woman earned only €0.02 per day. She was not financially able to buy the bamboo herself. Yunus decided to lend her and a number of other stool makers the money to buy the bamboo themselves, so that they would be able to set up their own company and build a better future for themselves.

At that time, no bank was prepared to issue any form of credit to these poor people. They were not able to hand over any assurance or security to a bank that could serve as a guarantee that the loan would be repaid. Now, various regular banks are prepared to issue similar microfinance. They have seen that it helps to improve the economic position of the poor and it has also earned the bank some money.

Box 6.6 Microfinance lending

Besides these forms of issuing credit, the position of the poorest communities can also be improved by developing products and services that have been adapted to their specific social circumstances, such as making products that help to combat certain illnesses (adding iodine to salt to prevent mental infirmities) or providing cost-efficient medical services (Prahalad 2005).

6.4 Conclusions

What conclusions can be drawn based on the above research concerning the way in which international companies can contribute to the local economy, particularly in developing countries? Views about this in the literature are too general and too ideological to offer companies a concrete guide on how to act. Therefore, an inventory has been taken of the practical experiences of representatives of nine companies with branches in developing countries.

The limited random survey of the nine companies showed that international companies contribute to the local economy in many different ways. This was endorsed by the interviewed representatives of social organisations and governments, although the social organisations, in particular, were suspicious of the actual contribution made by international companies. However, the doubts that they expressed concerning the input of international companies to strengthen the local economy were very general. This was because, as a result of a lack of information, these social organisations found it difficult to judge the actual input of companies on their merits. More transparency and openness from the companies should, in principle, solve this problem, because social organisations can then better examine the information provided by these companies.

An assessment instrument, which does not currently exist, is needed in order to assess the relevance of company information. A few companies, including Unilever (Oxfam/Unilever 2005), are busy developing an assessment instrument themselves. However, even though such an assessment instrument is desired, these efforts have not yet led to a more generally accepted instrument for assessing the advantages and disadvantages of foreign investment for the local economy of developing countries. The question that will arise is how extensive the information provided must be. Providing information quickly leads to the production of thick books that are not really easy for outsiders to understand. A solution would be to make a distinction between detailed information per country for the local inhabitants and generic information concerning all the countries where branches are located, which can be included in annual reports of corporate social responsibility.

Finally, all the parties (companies, social organisations and governments) agreed that international companies must primarily contribute to the local economy via their core activities. They can influence economic growth via taxation, wages and employment opportunities. International companies can also contribute to the economic development of the host country by supporting the physical infrastructure, transferring scientific and technological knowledge and making a positive contribution to local activities.

The importance of charity is recognised, but is considered to be less essential. Sponsoring, on the other hand, could be linked more firmly to the core activities of companies than is currently the case. Taking initiatives that can support the poorest communities (bottom of the pyramid) in improving their economic position is regarded as the most important challenge. In fact, this creates a new type of contribution towards the local economy: improving the position of the poorest communities by providing products and services that have been adapted to their specific situation. In this way, combating poverty, strengthening the local economy and strengthening the economic position of the company concerned go hand in hand. However, for international companies, applying the bottom-of-the-pyramid philosophy means a mental change, because the development of products and services must be adapted to the needs of the poorest communities. This requires fundamental technological changes and far-reaching empathy for the daily lives and activities of the communities, as well as empathy for the opportunities and obstacles experienced by the communities. If a company is successful in developing products and services that meet the needs of these people, this could be a breakthrough in reducing the gap between the rich and the poor.

7

The future of corporate social responsibility

7.1 There is not only one future

Nearing the end of this book, the question arises as to what the future of corporate social responsibility will be. Should companies take into consideration that more or less attention will be given to this theme? What attitude will various stakeholders take and what effect will this have on the international business sector? Which subjects within corporate social responsibility will remain prominent on the company agenda? And what will this mean for the reporting of companies concerning their performance on corporate social responsibility? A straightforward answer cannot be given to these questions. Much depends on the environment in which the international business sector will have to operate and the dominant values in this environment. For example, how will people view the free-market economy, regulation introduced by government, the globalisation of the economy and the co-ordinating power of international organisations, such as the United Nations?

In order to try to get an idea of the possible future of corporate social responsibility in an international context, four world-views are sketched below. These world-views have been drawn up by the Netherlands Environmental Assessment Agency (2004) for the National Institute for Public Health and the Environment (RIVM). The central question is: what consequences do these four scenarios have for the future of corporate social responsibility in an international context? The following will be discussed:

1. The attitude of business and stakeholders to corporate social responsibility in an international context

2. The themes to be addressed

3. The attention given to reporting

7.2 Four world-views

The Netherlands Environmental Assessment Agency (MNP) has distinguished four world-views:

A1: Global market

B1: Global solidarity

A2: Safe region

B2: Caring region

These can be considered as archetypes which exist in parallel and must not be interpreted too rigidly. People can sometimes identify with a number of world-views, but will generally have more affinity with one particular world-view.

In Figure 7.1 these four world-views are placed in a matrix. The vertical axis gives the degree of internationalisation of the activities (globalisation versus more regional development) and the horizontal axis

GLOBALISATION

**Global market
(A1)**

**Global solidarity
(B1)**

EFFICIENCY

SOLIDARITY

**Safe region
(A2)**

**Caring region
(B2)**

REGIONALISATION

FIGURE 7.1 Four world–views

Source: RIVM 2004

reflects the consideration between efficiency and solidarity and has to do with the choice between a free market and government co-ordination. Each world-view represents a different, specific quality of life. The Netherlands Environmental Assessment Agency explains the contents of the four world-views as follows:

7.2.1 Global market (A1)

Increasing globalisation and individualisation is the global market world-view's foundation. There is strong economic growth in this liberal, individualistic and efficiency-oriented world. The ecological risks (in particular, climate change) are large, as are the sociocultural risks (loss of social coherence, cultural identity and international solidarity). In this world-view, the world population will stabilise in the middle of the 21st century at 9 billion people. The world has a great deal of technological optimism.

7.2.2 Global solidarity (B1)

In this world-view oriented towards global solidarity, an attempt is made to steer globalisation in the right ecological and social direction through rules and treaties. An example of this is the Kyoto Protocol for tackling climatic problems. In this world-view, the institutions affiliated to the UN receive an increasing amount of legitimacy. There is effective worldwide government co-ordination, social justice is seen as an essential element for solving the tension between the economy and ecology (no intergenerational solidarity without international solidarity), and fighting hunger and poverty will be very important issues. The lower economic growth in Europe which is a consequence of this world-view is accepted.

7.2.3 Safe region (A2)

In the safe region world-view, the trend of hedonism and individualisation is continued. Supporters of this world-view are strongly oriented towards (national) safety, order and authority. Free trade is seen as a threat to employment opportunities. In the A2 world, modernisation will fail to materialise due to the exclusion of vulnerable regions. As a result, the total world population will increase to 11 billion in 2050. The very uneven division of wealth in this world-view will lead to a greater risk of global tension and conflicts. The A2 answer to this is less immigration and more security.

7.2.4 Caring region (B2)

Community spirit, civic duty and sociocultural diversity have a high priority in the caring region world-view. Immaterial issues, such as free time and social identity, are considered to be important in this world-view; money does not make you happy. The economic growth in this *small is beautiful* world-view is moderate (only 40% higher than in 2000) compared with the A1 world-view (140% higher). There is also a strong regional orientation, meaning products from your own local region and a large trust in local management (self-sufficiency).

7.3 Consequences of the four world-views for corporate social responsibility in an international context

What are the consequences of the four world-views described above for the interpretation of corporate social responsibility in an international context?[1] The insight gained during an expert meeting is summarised below for each world-view.

7.3.1 World-view A1: global market

The business sector is expected to have a defensive attitude in the global market world-view. An ad hoc approach will be followed by companies, aimed at maintaining their reputation. Matters will be implemented based on efficiency considerations. If it is more efficient

---

1 The question was addressed during an expert meeting with representatives of the companies involved in the programme 'CSR in an International Context', the National Institute for Public Health and the Environment (RIVM), Oxfam and two trendwatchers (Carl Rohde and Gijs ten Kate).

to move labour elsewhere, then this will be done. Chain responsibility will also be taken for reasons of efficiency and reputation management. Companies in international supply chains will work in accordance with legislation and transfer the risks to their suppliers.

The world-view can be considered to be the ultimate implementation of self-interest. This can go in one of two ways: corporate social responsibility no longer plays a role or a company distinguishes itself in the market through corporate social responsibility and, in this way, improves its reputation. In the latter case, this world-view has a more positive effect on corporate social responsibility and allows room for experimentation.

NGOs will mainly be concerned with large brands. They will probably be more critical of the defensive attitude of the business sector and their actions will mainly take advantage of the fear companies have of having their reputation damaged.

There will still be a link between the environment and economic growth. As a result of strong economic growth, a lot of attention is given to profit and much less attention is given to the environment (planet). Damage to the environment increases. Companies that wish to act proactively to get a good reputation can do so by improving their performance with regard to the environment. The business sector has great optimism in technology. Globalisation is increased as a result of the central position given to technology.

As a result of the emphasis placed on limiting damage to a company's reputation and acting in accordance with legislation, reporting on corporate social responsibility will take place within a closely defined context and will be based on a fixed set of indicators.

7.3.2 World-view B1: global solidarity

The emphasis in the global solidarity world-view will probably be placed on self-regulation by companies and the creation of partnerships and multi-stakeholder initiatives. The government's supporting role will be reduced as a result of such mutual agreements. The business sector uses a broader definition for the *creation of economic value*. There is more co-operation in international product chains than in the global market world-view and it is more natural to have agreements between the links in product chains. There are no front-runners in the field of corporate social responsibility. Since corporate social responsibility becomes a way of life, its position is not so special in this world-view.

The role of NGOs is to co-operate and negotiate with the business sector. The number of partnerships with the business sector (and also with the government) increases. The government will monitor observance of legislation more stringently. Since the expectations with regard to corporate social responsibility are high, companies are judged based on their performance in this field. Free-riders are tackled hard. As a result of the large amount of attention given to corporate social responsibility, the consumer seems to have been sidelined. In practice, the situation is probably a bit more subtle than that, because spirituality and corporate social responsibility are also important issues for consumers and stakeholders. Companies will take this into consideration, since they will still want to please the consumer.

In this world-view, it is more difficult for a company to make a name for itself with regard to corporate social responsibility, so the reporting of corporate social responsibility will probably become less important. Only the few free-riders will get a negative name for themselves. The large technological co-operation programmes between the government and the business sector will also receive the attention they require.

7.3.3 World–view A2: safe region

It is to be expected that corporate social responsibility in the safe region world-view will be strongly oriented towards the local community. Protecting your own market will be a central theme. Companies will mainly be concerned with strengthening their local identity (therefore, local brands). Central concepts are conflict, survival, short term and security. There are few possibilities for front-runners to make a good name for themselves with regard to corporate social responsibility, except with local themes. Companies place emphasis on acting in accordance with their local government legislation. Regional issues will occupy a central position in the approach of multinational companies. They must adapt their strategy to the local conditions and identity and, when doing so, take cultural and local differences into consideration. Opinions within one region can often also clash. This requires extra attention, because companies will have to make many different parties happy. This is why multinationals will be inclined to limit themselves to meeting regional demands. The involvement of stakeholders is probably high, because companies must be in discussion with local NGOs and other local stakeholders.

There will probably be two types of NGO: traditional NGOs (often globally oriented and placing emphasis on problems caused by the Western economies passing the buck to developing countries) and local NGOs (who urge companies to do good for their own region). The latter group of NGOs is a kind of 'not in my back yard' movement, but dressed up differently. These more locally oriented interest groups will be less oriented towards global themes. The liability approach (following the American model) will be a main NGO strategy issue.

The themes addressed will mainly be the areas of consumer safety and the security of the local region. There is less support for corporate social responsibility in an international context, but more so for local corporate social responsibility. Technological innovations will also be aimed at local themes and safety.

Reporting with regard to corporate social responsibility from a central level is less important in this world-view. The emphasis lies with local reporting.

7.3.4 World-view B2: caring region

It is expected that corporate social responsibility will be aimed at the local environment in the caring region world-view, but not from a point of view of conflict as in the safe region world-view. The intention is to give heavy consideration to corporate social responsibility, because the theme is highly valued in society. Corporate social responsibility serves as a guiding principle for how companies act. The business sector has strong roots in society. The maxim is: allow my region to flourish and allow it to keep its characteristics. The mentality is: let's make something nice out of it. Chain responsibility in an international context is less strongly developed, but still receives some attention. People want to do something good for the poor people in the world. Local brands are central in this world-view. Product differentiation is, therefore, necessary to reach the local market.

Besides the traditional, broadly oriented NGOs, there will be many local, specialised and idealistic NGOs that will want to promote products from the regional market. Niche products with high social and environmental standards will have a much larger market share.

The priority will be the development of local products and anticipating local preferences. There is little room for innovation and there is little demand for technical innovations.

In this world-view, reporting on corporate social responsibility will not be as important. Aspects of sustainability will be integral to the production of local products.

7.4 Conclusions

From the tentative interpretation of the four world-views described above, it can be seen how different the attitudes towards corporate social responsibility can be. Each of the four world-views represents a specific quality of life. The world-views differ in the extent of the international interweaving of activities (globalisation versus more regional development) and in the extent of efficiency and solidarity (in combination with the choice between free market and government co-ordination). The attitude of businesses and stakeholders is different in every world-view. The emphasis is also placed on specific subjects

and, as a result, the reporting is different. This is why there is not only one future for corporate social responsibility in an international context. How companies accept their corporate social responsibility depends on the developments that occur locally and internationally, but it is beyond dispute that corporate social responsibility plays a role in every world-view.

8

Ten key practical experiences

This book has addressed a number of crucial issues which will confront companies wishing to implement corporate social responsibility in a global world. For instance, how can they find their way through the maze of regulations and standards that have been developed in the course of time? How can companies best deal with the tension between observing international rules of conduct and circumstances specific to the location? How can they be socially responsible for the product chain(s) in which they operate? And how can companies best contribute to developing countries' local economies?

The book is based on the experiences of 20 large and smaller companies which operate globally. The ten key practical experiences of this group of companies are summarised below.

1. Seeing the wood for the trees

Through the years, a maze of international guidelines and standards has been developed regarding corporate social responsibility. The following plan of action is useful to create some form of order in this maze:

- Action 1: assess the business's current state of affairs regarding corporate social responsibility in an international context based on the OECD *Guidelines for Multinational Enterprises*, which are the most encompassing guidelines. Determine priorities and a code of conduct based on this assessment. Communicate these priorities and the code of conduct to the stakeholders

- Action 2: develop the policy priorities using theme-specific international guidelines and standards concerning corporate social responsibility

2. A code of conduct forms the core of the company's policy with regard to corporate social responsibility. At the same time, it is also a useful means to internally and externally communicate the international company policy with regard to corporate social responsibility

Internal communication is crucial to ensure that everybody in the organisation familiarise themselves with the contents of the code of conduct. Different methods of communication must be used to explain the contents of the code of conduct, why it is important and what top management wishes to achieve with it. In order to gain external support for the code of conduct, it is important to engage in dialogue with the stakeholders concerning the code at an early stage. This makes it possible to include their views in the considerations and it will also increase acceptance of the choices to be made. However, practice has shown that a code of conduct contains more general wording than a company uses in practice. It is, therefore, also a good idea to give insight into the procedures and regulations the company has drawn up in order to support the internal decision-making process.

3. Moral and cultural considerations play a more important role for social themes than they do for environmental themes

Problems arise, for example, within areas such as free trade unions, discrimination and equal opportunities (men–women, religion, sexual orientation), working hours, the protection of privacy and rules concerning gifts. In such cases, it is not possible for the central office to outline in detail which rules local branches must observe. In order to ensure uniformity in the policy, companies can place demands on the internal process which must take place at local branches. This can be achieved by letting local managers indicate which dilemmas they are confronted with and how they wish to solve them. Through consultation with people at central level, it can be determined whether the proposals are in keeping with the general company policy. By maintaining clear and unequivocal procedures, companies can find a way of dealing with the tension between observing international rules of conduct and local circumstances.

4. Internal company factors are very important when determining whether the general company policy will be successfully implemented at different local branches with regard to corporate social responsibility

The local manager's vision and his or her personal style of leadership appear, in practice, to be important keys to success. The place occupied by social and environmental policies within the organisational structure (in particular the authority of people responsible) is crucial. Finally, the company management team's communication style also plays a role. It must be in keeping with the branch culture and the most effective mix of communication methods differs by country.

5. A plan of action is necessary in order to implement corporate social responsibility at all branches throughout the world

The plan of action shown in Box 8.1 has been drawn up based on company experiences.

6. The business strategy regarding corporate social responsibility should be attuned to the societal needs and customs of the country concerned

In view of the large differences in social context between different countries, companies that operate internationally must take the political context of the country concerned, as well as the social problems and customs, into consideration when determining their policy concerning corporate social responsibility. It is sensible for a company to adapt its strategy to the local circumstances of the country in which it invests, but without disregarding its own standards and values.

7. The organisation of chain responsibility in an international context demands a structured approach

Based on company experiences, a step-by-step plan has been developed to help companies give shape to chain responsibility in an international context. The guiding principles of this step-by-step plan are the differences in the following four variables: the complexity in the chain, the level of ambition, the diversity of the product range and the power the company has in the chain. A summary of this step-by-step plan is given in Box 8.2.

1. Assess the current situation with regard to corporate social responsibility within the company and its branches and any relevant chain partners based on the OECD guidelines for multinational companies

2. Formulate a preliminary vision and code of conduct regarding corporate social responsibility based on the results of Step 1

3. Enter into dialogue with relevant local and international, as well as internal and external, stakeholders about their expectations and demands and reformulate the preliminary vision and code of conduct based on the findings

4. Develop short-term and longer-term strategies regarding corporate social responsibility in an international context and use them to draft a plan of action. Make use of theme-specific guidelines and standards

5. Integrate the approach chosen as much as possible into existing company processes and procedures. Set up a monitoring and reporting system. Make use of indicators to assess the progress made. Take into account the location-specific interpretation of corporate social responsibility within a centrally determined bandwidth. Draw up specific procedures and support documentation for this

6. Embed the process by incorporating the chosen approach into the quality and management systems. Assign responsibilities and tasks to departments and personnel and integrate these into the existing remuneration systems. Turn the approach into a continuous process of improvement

7. Communicate the approach and the results obtained internally and externally. Adapt the style of communication to the culture of the company branch and the local, external stakeholders. Ensure there is visible commitment from senior management

Box 8.1 Plan of action for implementing corporate social responsibility in an international context

Step 1. Determine those parts of the product chain for which you can/want to take social responsibility, taking into account the complexity of the chain.

Step 2. Do you focus the product chain on:

(a) a niche market?

(b) a mainstream market?

If a, then you can immediately adopt high environmental and social standards.

If b, then it would be better to introduce environmental and social standards which will become stricter over time.

Step 3. Determine the degree of diversity in the product assortment.

(a) Do you purchase a great diversity of products?

(b) Are you part of one or a limited number of product chains?

If a, go to Step 4.

If b, go to Step 5.

Step 4. Develop a strategy to ensure that your most important suppliers comply with your ethical code of conduct.

Step 5. Are you able (perhaps in co-operation with a limited number of influential actors) to impose standards on your suppliers and/or other companies in the product chain?

(a) Yes

(b) No

If a, then you can take the initiative to organise chain responsibility.

If b, then, preferably, try to seek alliance with an influential organisation or competitors in the same sector who can help organise chain responsibility. If this is not possible, select a limited number of important suppliers with whom co-operation can be established.

Box 8.2 Step-by-step plan for global chain responsibility

8. There must be transparency in showing how observance of the international code of conduct concerning corporate social responsibility is guaranteed in the chain

In order to promote transparency it is desirable for the following items to be reported on:

1. The quality of and the method used by the audit company (including the involvement of the local environment and employees)

2. The results of the audit and the proposed corrective action (including a time schedule for when this action must have occurred)

3. The points where the policy deviates from international guidelines and standards and why that is the case

4. The way in which purchasers support their suppliers, either financially or otherwise, in making it possible for them to meet the requirements with regard to corporate social responsibility

9. Internationally operating companies can best contribute to the local economy via their core activities

Companies that operate internationally can contribute to improving the local economy, in particular of developing countries, in two different ways. They can contribute via charity activities or via their core activities. Since the latter path is of a more structural nature, this is generally the preferred path. Companies can influence economic growth and development of the host country via their core activities. Sponsoring can also be used to make a contribution. It is a challenge to take initiatives that can help the poorest communities at the bottom of the pyramid to improve their economic position. In fact, this creates a new type of contribution towards the local economy: strengthening the position of the poorest people by providing products and services that have been adapted to their specific situation. However, this requires fundamental technological changes and far-reaching empathy for the daily lives and activities of these people and the opportunities and obstacles they are faced with.

A widely accepted assessment framework is necessary in order to increase transparency with regard to the contribution that international companies make to the local economy of developing countries. Such a framework is still missing.

10. There is not only one future for corporate social responsibility in an international context

There is not only one future for corporate social responsibility in an international context. A tentative interpretation of four different world-views (global market, global solidarity, safe region and caring region) shows how different they can be. The attitude of businesses and stakeholders is different in every world-view. The emphasis is also placed on specific subjects and, as a result, the coverage is different. The way in which companies determine their strategy for corporate social responsibility in an international context, therefore, depends on the developments that occur both locally and internationally. It is beyond dispute, though, that corporate social responsibility plays a role in every world-view and it is, therefore, a good idea for companies to anticipate this. Companies will be able to do this using the practical experiences given in this book. This will help to ensure that corporate social responsibility in a global world is not seen as a threat, but rather as an opportunity.

APPENDIX 1

The 'Corporate Social Responsibility in an International Context' programme

The book is based on the experiences of the following 20 companies:

1. ABN AMRO (bank)

2. AgroFair (producer of certified fruits)

3. AXA Stenman Holland (producer of locks for windows, doors and bicycles)

4. De Bijenkorf (large retailer)

5. Bloemenveiling Aalsmeer (flower auction)

6. The Cookie Company (producer of licensed baby and children's products)

7. Difrax (producer of baby products)

8. Friesland Foods (company focusing on development, production and sale of a wide range of dairy products and fruit drinks)

9. Fugro (company focusing on collecting, processing and interpreting data about the Earth's surface and the seabed)

10. Heineken (company focusing on production, distribution and sale of beer, malt and soft drinks)

11. Koninklijke Houthandel G. Wijma & Zonen (Royal Wijma timber trading company)

12. Koninklijke Wessanen (Royal Wessanen; food business group)

13. Koninklijke KPN (Royal KPN; telecommunications company)

14. Merison (supplier of non-food products to supermarkets)

15. Pentascope (consultancy firm which implements changes in organisations)

16. Royal Haskoning (engineering consultancy firm)

17. Rijnvallei (cattle feed producer)

18. Shell (energy and petrochemicals producer)

19. Simon Lévelt (coffee supplier)

20. Thermphos (manufacturer of phosphorus, phosphorus derivatives, phosphoric acid and phosphates)

These companies, which participated in the 'Corporate Social Responsibility in an International Context' programme, represent a wide variety of sectors and they also vary in size.[1] The programme was

---

1 This programme is the follow-up to the 'From Financial to Sustainable Profits' programme, which ran from 2000 to 2002 within the framework of the National Initiative for Sustainable Development (NIDO). The results of this programme have been published in two books: Cramer 2002, 2003.

carried out by CSR Netherlands (previously the National Initiative for Sustainable Development [NIDO]) and ran from the beginning of 2003 to the end of 2005. The programme was financed by the Dutch Ministry of Housing, Regional Development and the Environment.

The objective of the programme was to make clear how an international company can show corporate social responsibility, in various countries, while taking the local culture, the social context and the local government's policy into consideration. Moreover, the programme outlined how companies can accept their corporate social responsibility in the international chains in which they operate.

Learning from each other's experiences was the programme's main issue. Companies exchanged experiences with each other during monthly meetings. Where necessary, the programme management supported the process by contributing specific scientific knowledge themselves or by including third parties. The participating companies also engaged in dialogue with various groups of stakeholders, which led to their vision and approach being tested and more clearly defined. The companies then used the experiences gained in their own work situations. The companies also all carried out their own company project during the course of the programme. This project showed common ground with the central themes of the programme.

How far advanced the participating companies were with implementing corporate social responsibility in an international context varied greatly. A few were front-runners, the majority were reasonably advanced and some were only just starting out. The reasons for becoming actively involved in this theme were equally diverse. One company saw it primarily as a market opportunity, another company put its reputation first and, for a third company, increasing employee motivation was the most important reason. Despite these differences, the experiences of the participating companies with implementing corporate social responsibility in an international context were similar. All the companies found it to be a process of searching; each company had to determine its own priorities. There is no standard concept for giving shape to corporate social responsibility in an international context and for many companies it is, therefore, a case of learning as you go.

APPENDIX 2
Main guidelines and standards for international corporate responsibility

(see Chapter 2)

Human rights

The Universal Declaration of Human Rights

This is the framework for human rights of the United Nations. It concerns a performance-oriented standard (www.udhr.org). The standard 'Norms on the Responsibilities of Transnational Corporations and Other Business Enterprises with Regard to Human Rights' is under construction, but as yet consensus has not yet been reached within the UN.

Website: www.un.org/Overview/rights.html

Labour rights

The International Labour Organisation: Tripartite Declaration of Principles concerning Multinational Enterprises and Social Policy

This is an integration of the most important ILO conventions and recommendations. It is a performance-oriented standard.
Website: www.ilo.org/public/english/standards/norm/sources/mne.htm

Social Accountability 8000

This is a worldwide standard in the area of labour rights. It is a performance- and process-oriented standard with possibilities for certification and a certifiable standard.
Website: www.sa–intl.org

OHSAS 18001

This is a process-oriented standard for policies in the health and safety area.
Website: www.ohsas.org

AccountAbility 1000 Framework

Process-oriented standard for social and ethical accounting
Website: www.accountability.org.uk/aa1000/default.asp

Environment

The Rio Declaration on Environment and Development

This is the UN starting point regarding the environment and development. Performance-oriented standard.
Website: www.un.org/documents/ga/conf151/aconf15126–1annex1.htm

The CERES (Coalition for Environmentally Responsible Economies) principles

Ten performance-oriented and, to a lesser extent, process-oriented principles which cover the most important environmental issues and also form the basis for the GRI guidelines.

Website: www.ceres.org/coalitionandcompanies/principles.php

The Natural Step

Comparable with the CERES principles: performance-oriented and, to a lesser extent, process-oriented.

Website: www.naturalstep.org

ISO 14001

Much used, process-oriented international standard for an environmental management system.

Website: www.iso.org

Corruption

The OECD Convention on Combating Bribery of Foreign Public Officials in International Business Transactions

Legal framework for coping with corruption practices. This is a legally binding performance standard.

Website: www.oecd.org/document/21/0,2340,en_2649_34859_2017813_1 _1_1_1,00.html

The Business Principles for Countering Bribery

Multi-stakeholder framework concerning corruption; both performance- and process-oriented guidelines.

Website: www.transparency.org/building_coalitions/private_sector/ business_principles.html

Economy

Guidelines and standards related to the company's contribution to economic welfare in the broadest sense are lacking (so far).

Generic

See the OECD guidelines and the UN Global Compact.
Websites: www.oecd.org/document/28/0,2340,en_2649_34889_2397532 _1_1_1_1,00.html
www.unglobalcompact.org

In addition to the OECD guidelines for multinational enterprises and the UN Global Compact, the following generic guidelines may also be helpful:

IFC guidelines of the World Bank

These performance-oriented guidelines provide criteria for financing projects in the area of social and environmental policy.
Websites: www.ifc.org   www.ifc.org/enviro
www.ifc.org/ifcext/enviro.nsf/Content/EnvironmentalGuidelines; the guidelines described here are sector-specific.

The 'Sustainability: Integrated Guidelines for Management' (SIGMA) project

An attempt to integrate all aspects of corporate social responsibility in one framework; this process standard is not yet generally accepted and is still under preparation.
Website: www.projectsigma.co.uk/Guidelines

Global Reporting Initiative

Process-oriented guidelines for the standardisation of reporting methods for corporate social responsibility; based on the main guidelines and standards; increasingly accepted as the basis for reporting
Website: www.globalreporting.org

Abbreviations

AIDS	acquired immuno-deficiency syndrome
BSCI	Business Social Compliance Initiative
CEO	chief executive officer
CRA	Country Risk Assessment
ESHIA	Environmental Social Health Impact Assessment
ESI	Environmental Sustainability Index
FIDIC	International Federation of Consulting Engineers
FFP	Fair Flowers and Plants
FSC	Forest Stewardship Council
GRI	Global Reporting Initiative
HIV	human immunodeficiency virus
HRCA	Human Rights Compliance Assessment
ICTI	International Council for Toy Industries
ILO	International Labour Organisation
ISO	International Organisation for Standardisation
MNP	Environmental Assessment Agency (Netherlands)
MPS	Milieu Programma Sierteelt (Environmental Programme for Floriculture)
NGO	non-governmental organisation
NIDO	National Initiative for Sustainable Development (Netherlands)
OECD	Organisation for Economic Co-operation and Development
RIVM	National Institute for Public Health and the Environment (Netherlands)
SWOT	strengths–weaknesses–opportunities–threats
UN	United Nations

References

Barrez, D. (2001) *De Antwoorden van het Antiglobalisme: Van Seattle tot Porto Alegre* (*The Answers to Anti-Capitalism: From Seattle to Porto Alegre*) (Alphen aan den Rijn, Netherlands: Haasbeek).

Cavanagh, J., and J. Mander (2002) *Alternatives to Economic Globalization: A Better World is Possible* (a report of the International Forum on Globalisation; San Francisco: Berrett-Koehler): 5.

Corporate Plaza (2005) 'Ondernemen met hoofd en hart: Duurzaam ondernemen; praktÿkervaringen', *Corporate Plaza* 1.

Cramer, J. (2002) *Entrepreneurship with Head and Heart. Corporate Social Responsibility: Practical Experiences* (Assen, Netherlands: Koninklijke van Gorcum).

—— (2003) *Learning about Corporate Social Responsibility: The Dutch Experience* (Amsterdam: IOS Press).

—— and F. Klein (2005) 'Ketenverantwoordelijkheid staat nog in de kinderschoenen' ('Chain Responsibility is Still in its Infancy'), *Arena* 3: 14-15.

Forrester, V. (2001) *De Terreur van de Globalisering* (*The Terror of Globalization*) (Amsterdam: Rainbow Pocketboeken).

Fortanier, F. (2004) 'The Impact of Foreign Direct Investment on Sustainable Development: Reviewing the Evidence', in J. Cramer, F. Engering, F. Fortanier, M. Hillen, Z. Oueslati, R. Porte and M. Smit (eds.), *Investing in Developing Countries: The Future Role of FDI* (The Hague: SMO): 23-35.

GRI (Global Reporting Initiative) (2005) *GRI Boundary Protocol* (Pilot version; Amsterdam: GRI).

Hart, S.L. (2005) *Capitalism at the Crossroads: The Unlimited Business Opportunities in Solving the World's Most Difficult Problems* (Upper Saddle River, NJ: Wharton).

Hertz, N. (2002) *De Stille Overname: De Globalisering en het Einde van de Democratie* (*The Silent Takeover: Globalisation and the End of Democracy*) (Amsterdam: Contact).

ICC (International Chamber of Commerce) (2004) *Standing Up for the Global Economy: Key Facts, Figures and Arguments in Support of Globalisation* (Paris: ICC, June 2004; www.iccwbo.org).

Jenkins, R., R. Pearson and G. Seyfang (2002) *Corporate Responsibility and Labour Rights: Codes of Conduct in the Global Economy* (London: Earthscan).

Klein, N. (2000) *No Logo* (London: Flamingo).

Leipziger, D. (2003) *The Corporate Responsibility Code Book* (Sheffield, UK: Greenleaf Publishing).

Mamic, I. (2004) *Implementing Codes of Conduct: How Businesses Manage Social Performance in Global Supply Chains* (Sheffield, UK: Greenleaf Publishing/ ILO).

Naert, P., and B. Coppieters (2000) *Globalisering: Zegen en Vloek* (*Globalisation, Blessing and Curse*) (Tielt, Belgium: Lannoo).

Neef, D. (2004) *The Supply Chain Imperative: How to Ensure Ethical Behaviour in Your Global Suppliers* (New York: AMACOM).

Netherlands Environmental Assessment Agency (2004) *Quality and the Future: Sustainability Outlook* (Bilthoven, Netherlands: RIVM, SDU).

New, S., and R. Westbrook (2004) *Understanding Supply Chains: Concepts, Critiques and Futures* (Oxford, UK: Oxford University Press).

Norberg, J. (2002) *Leve de Globalisering* (*Long Live Globalisation*) (Antwerpen, Belgium: Houtekiet).

OECD (Organisation for Economic Co-operation and Development) (2000) *The OECD Guidelines for Multinational Enterprises* (Paris: OECD; www.oecd.org).

Oxfam/Unilever (2005) 'Exploring the Links between International Business and Poverty Reduction: A Case Study of Unilever in Indonesia', downloadable at www.novib.nl.

Porte, R., and J. Cramer (2005) 'The Contribution of Multinational Organisations to the Local Economy of Developing Countries', in J. Cramer, F. Engering, F. Fortanier, M. Hillen, Z. Oueslati, R. Porte and M. Smit (eds.), *Investing in Developing Countries: The Future Role of FDI* (The Hague: SMO): 37-51.

Prahalad, C.K. (2005) *The Fortune at the Bottom of the Pyramid: Eradicating Poverty through Profits* (Upper Saddle River, NJ: Wharton).

Soros, G. (2002) *Globalisering* (*Globalisation*) (Amsterdam: Contact).

United Nations (2000) *UN Global Compact's Ten Principles* (New York: UN; www.unglobalcompact.org).

Van Liemt, G. (2002) *Het Wereldbeeld van Antiglobalisten* (*The World View of Anti-Globalists*) (The Hague: SMO).

World Bank (2002) *Globalization, Growth, and Poverty: Building an Inclusive World Economy* (World Bank Policy Research Report; Washington, DC: World Bank).

Index